On Evolution

On Evolution

John C. Avise

The Johns Hopkins University Press | Baltimore

9 8 7 6 5 4 3 2 1

The Johns Hopkins University Press
2715 North Charles Street
Baltimore, Maryland 21218-4363
www.press.jhu.edu

Library of Congress Cataloging-in-Publication Data
Avise, John C.
    On evolution / John C. Avise
    p. cm.
    Includes bibliographical references and index.
    ISBN 978-0-8018-8688-1 (hardcover : alk. paper)
    ISBN 978-0-8018-8689-8 (pbk. : alk. paper)
    1. Evolution (Biology) I. Title.
    QH366.2.A95 2007
    576.8—dc22   2007006283

A catalog record for this book is available from the British Library.

Contents

Publisher's Note

Long a friend, and longer a fan, of John Avise, I approached him in 2005 to see whether he would be willing to revise some of his most interesting writings to create a book "on evolution." His clear, accessible writing style and his impeccable credentials—distinguished professor, member of the National Academy of Sciences, founder of phylogeography, author of numerous books and innumerable papers—made him, I reasoned, an ideal author for such a task. His insights, presented here with passion but without pretense, make complicated topics understandable. All of the pieces included have been rewritten and edited to provide the reader with an accessible look into the field of evolutionary biology. Insightful readers will also be able to follow the arc of evolutionary biology over the past three decades, from the early illumination provided by protein electrophoresis to the recent ability to scan entire genomes. Through it all, John Avise guides us with an unwavering hand, never losing sight of the fact that it's not just molecules we are talking about, it's also nature. These writings are indispensable reading for all who have an interest in evolution.

Vincent Burke, Senior Editor
The Johns Hopkins University Press

Preface

When my editor approached me about transforming some of my former writings into this volume, I agreed, but with trepidation. Evolution is a huge topic, and I worried that creating a readable book using revisions of merely a small handful of my works would not do the field justice. Indeed it cannot. But a short book that deals broadly with evolution and its by-product, biodiversity, seemed important enough for the effort.

Editorial stipulations were that the collection should include papers spanning the four decades of my career, that these works should canvass a wide variety of biological topics, and, in fairness to others, that they should emphasize work in which I took a lead role (despite the fact that more than 80% of the publications from my laboratory have involved shared authorship with graduate students or other colleagues). The intent would be to produce a work that is both useful and entertaining to three audiences: professional biologists, students, and intelligent nonscientists.

I have revised and shortened most of the included works and have kept a few lighter pieces that one can argue are tangential to evolution. I hope these lighter-tone pieces counterbalance the few that might require extra effort on the part of the nonscientist. Some of the pieces I chose to include may make my professional colleagues wince a bit, but overall I hope that scientists and students and others will find this book fun as well as illuminating.

I struggled with how to arrange this book, but eventually decided that the pieces should travel through time. The first articles (written in the 1970s) required substantial rewriting to make them relevant, yet in the end I think they retain the original elements that give them a certain charm. The latter pieces in general required less revision. A few selections address admittedly difficult, but far from impenetrable, subjects. In the end, I hope I have produced a book that provides an enjoyable journey into the wide realm of evolutionary biology, at least along the sundry paths that I have traveled.

On Evolution

1

Genetic Differentiation during Speciation

Ever since Darwin, a central issue in evolutionary biology has been whether closely related species differ substantially or only trivially in their genetic features. In the 1960s, molecular genetic techniques (notably protein electrophoresis) were introduced to population biology, and these procedures gave new opportunities to examine the topic of speciation empirically. This paper, written while Avise was a graduate student in Francisco Ayala's laboratory at the University of California at Davis, addresses two allied but subtly distinct evolutionary questions that were important in the 1970s and remain so today: What molecular genetic changes accompany the formation of new species, and what genetic changes are actually responsible for the origins of reproductive isolation? Readers wishing a superb update on speciation topics discussed in this early review should consult J. A. Coyne and H. A. Orr's *Speciation* (2004).

Biological evolution consists of two processes: anagenesis (or phyletic evolution) and cladogenesis (i.e., splitting). Anagenetic change is gradual and usually results from increasing adaptation to the environment. A favorable mutation or other genetic change arising in a single individual may spread to all descendants by natural selection. Cladogenesis results in the formation of independent evolutionary lineages. Favorable genetic changes arising in one lineage cannot spread to members of other lineages. Cladogenesis is responsible for the great diversity of the biological world, allowing adaptation to the variety of ways of life. The most decisive cladogenetic process is speciation.

Among sexually reproducing organisms, species are groups of interbreeding natural populations that are reproductively isolated from other such groups. Gene exchange can occur among Mendelian populations of the same species. The speciation process requires the development of reproductive isolation between populations, resulting in independent gene pools. Two related questions concerning speciation interest evolutionists: What ecological and evolutionary conditions promote speciation, and what changes in the genetic composition of populations result in reproductive isolation?

For sexually reproducing organisms, isolation by geographic barriers and the concomitant severe restriction of gene exchange is the usual prerequisite to genetic divergence and speciation. Geographically isolated populations accumulate genetic differences as they adapt to their different environments (or sometimes as they merely drift apart in genetic composition). In the short run, they may become recognizable as races. However, not all races will become species because the process of geographic differentiation is reversible. If the races have not sufficiently diverged while separated (allopatric), they may later converge or fuse through hybridization. On the other hand, allopatric populations may sometimes become sufficiently different genetically, so that if the opportunity for gene exchange ensues again, hybrids will have low fitness. Natural selection would, then, favor the completion of reproductive isolation.

Geographic Speciation

Two stages may be recognized in the process of geographic speciation. During the first stage, populations become isolated by geographic barriers and accumulate genetic differences. Much of this divergence is the result of adaptation to different environments, but other factors such as genetic drift and founding events may play a role. Partial or even complete reproductive isolation between populations may develop as a by-product of this genetic divergence. During the second stage of speciation, natural selection may hasten the direct development of reproductive isolation in the form of prezygotic isolating barriers. This stage begins when genetically differentiated populations regain geographic contact. If reproductive isolation is not yet complete and if the gene pools have sufficiently diverged, matings between individuals of different populations may produce progenies of reduced quality (thus, lower fitness). Natural selection would then favor genetic variants that promote matings between members of the same population. Reproductive isolation would thereby be enhanced. Two survey strategies have been employed in attempts to determine the degree of genetic differentiation during speciation. A direct strategy involves assaying populations that appear to be in various stages of the speciation process. Such studies permit assessment of the amount of genetic differentiation during the first stage of speciation when allopatric populations develop incipient reproductive isolation, and during the second stage of speciation when reproductive isolation is being completed by natural selection between populations that have reconnected (regained sympatry). A second survey strategy involves assaying populations belonging to different species. Species' differences represent the sum of genetic differences accumulated subsequent to speciation as well as during the speciation process itself. Hence, interest has centered on species that by other criteria appear particularly closely related, such as morphologically similar species (sibling species) and species that can hybridize.

In a classic study of genetic differentiation during geographic speciation, Ayala and colleagues examined the *Drosophila willistoni* complex, which consists of more than a dozen closely related species endemic to the American tropics. Some species, such as *D. nebulosa,* are readily distinguishable morphologically from the rest. Other species are called siblings because they are morphologically nearly indistinguishable. Despite their morphological similarity, sibling species are completely isolated reproductively.

The fruit fly populations that Ayala studied were at five increasing levels of evolutionary divergence: geographic populations within a taxon; subspecies, in the first stage of geographic speciation; semi-species, in a second stage of speciation; sibling species, in which speciation is complete but little morphological divergence has accumulated; and nonsibling species, exhibiting morphological differences as well as reproductive isolation. The sibling species in this case could only be distinguished morphologically by minute differences in male genitalia, yet genetically they proved to be very different from one another.

The major conclusions obtained in the Ayala studies were that (1) very little genetic differentiation exists between local populations within a species; (2) populations showing incipient reproductive isolation, often in the form of hybrid sterility, exhibit significantly greater genetic distances, involving molecular changes in at least 20% of their structural genes; and (3) populations between which reproductive isolation, and therefore speciation, is complete usually display distinct forms (alleles) at one-third or more genetic loci. These observations strongly support the contention that numerous genetic changes often accompany the speciation process.

Beyond Ayala's classic work, many studies exist on genetic differentiation between vertebrate populations in early stages of evolutionary divergence. The sunfish genus *Lepomis* contains eleven species, all native to North America. These species are renowned for their ability to hybridize, especially in disturbed ecological conditions or in the laboratory. First-generation (F_1) hybrids from twenty-one different combinations of two species have been found in nature. Nonetheless, the various species retain their identities throughout their ranges, even where they are sympatric. Adults of all species can readily be distinguished morphologically. Despite their ability to hybridize, *Lepomis* species proved to be very distinct genetically—about 50% of their genetic loci show different allelic forms in a typical comparison between two species.

Turning to amphibians, the salamander genus *Taricha* consists of three species: *T. granulosa, T. rivularis,* and *T. torosa.* Two allopatric subspecies of *T. torosa* are recognized: *T. t. torosa* and *T. t. sierrae.* Reproductive isolation between *Taricha* species is of a partly different nature than that in *Drosophila.* There is no obvious physiological reason why the various *Taricha* species cannot reproduce; reproductive isolation is apparently maintained solely by behavioral barriers to mating. Nonetheless, the magnitude of genetic difference between salamander species turned out to be similar to that observed between the *Drosophila* species, and the sunfish species. Two subspecies of the house mouse (*Mus mus musculus* and *M. m. domesticus*) are largely allopatric in Europe, but they do meet and hybridize along a boundary running east to west through central Denmark. Reproductive isolation appears to be maintained through reduced fitness of backcross progeny due to disruption of coadapted parental gene complexes, perhaps coupled with environmental differences favoring different genetic compositions on either side of the hybrid zone. The genetic differences between these two forms of mice again proved to be large and comparable to those between closely related species of fruit flies, sunfishes, and salamanders.

Although geographic speciation is the most common mode of speciation among sexually reproducing outcrossing animals, speciations may also occur through rapid, initially nonadaptive means. Such *saltational* speciation events may involve polyploidization (changes in the number of chromosome sets), rapid chromosomal reorganizations, or changes in breeding system. In such cases, speciation may be completed with little or no change at the genic level. For example, new autopolyploid or allopolyploid species contain no new alleles not already present in their ancestors. Polyploidization and changes in breeding system are rare among higher animals, but these events do play important roles in speciation of many plants. However, one mode of saltational speciation—rapid chromosomal reorganization —does seem to be common in some groups of higher animals as well.

Significance of Structural Gene Changes during Speciation

If one considers the grossly different processes that may be involved in the speciation of organisms as different as fruit flies, fishes, salamanders, and mammals, the range of genetic distance estimates among subspecies, semispecies, and closely related species is surprisingly small. When compared with levels of genetic differentiation among local populations, these estimates suggest that a substantial proportion of genes may be changed in allelic composition concomitant to the speciation process. Typically, about twenty electrophoretically detectable allelic substitutions per 100 loci accumulate before reproductive isolation is completed. Arguments that speciation is normally accompanied by little genic change are simply not substantiated by the evidence.

Estimates of genetic divergence between closely related species support the contention that many genic changes occur during speciation. Even species that appear closely related by other evidence, such as morphological similarity or hybridizing propensity, are often distinct in allelic composition at one-fourth to one-half of their loci. Populations continue to accumulate genetic differences following speciation. By the time they have diverged sufficiently to warrant their placement in different genera by conventional systemic criteria, they often share common alleles at only a minority of loci.

Nonetheless, the range in biochemical similarities between related species is large, and some species appear little or no more distinct than do local populations within a species. Taken at face value, these instances argue that not all speciations entail large genic changes. As well, not all speciation events involve the same amount of genetic differentiation, even when reproductive isolation arises according to the conventional model of geographic speciation.

Evidence from fruit flies suggests that substantial proportions of genes are changed during the first stage of geographic speciation, but that few additional changes occur during the second stage when the development of reproductive isolation by natural selection is taking place. Perhaps the number of genes required to develop reproductive isolation per se is small. We would then expect that, in situations where natural selection favors the development of reproductive isolation (such as between populations differing in chromosome numbers or arrangements), change in only a few genes could complete the speciations. If this is true, and if the diverging populations that differ in chromosomal content were initially similar at the single-gene level, then at the outset of reproductive isolation the two emerged species would show only small genetic distances. This line of reasoning is also consistent with the finding of similar levels of genetic differentiation in species-rich versus species-poor groups of fish of about equal evolutionary age, which suggests that time is a primary determinant of genetic differentiation. During the first stage of speciation, time may be an important predictor of the accumulation of genetic differences as populations adapt to their environments. The development of reproductive isolation per se may not substantially increase the level of genetic divergence.

This raises the question of the relevance of structural gene divergence to speciation. Although a large number of structural gene changes normally precede the completion of reproductive isolation, it is conceivable that these allelic substitutions are irrelevant to the development of reproductive isolation. Some authors have argued that structural gene evolution is primarily a function of mutation rates to neutral alleles. By definition, neutral alleles confer identical fitnesses on their bearers and, hence, could not contribute to the development of reproductive isolation. On the other hand, if structural genes do affect fitness, depending on the genetic and environmental backgrounds in which they occur, they may play an important role in speciation. Perhaps the structural gene differences that accumulate during the first stage of geographic speciation contribute to the loss of hybridizing ability or to the lowered fitness of hybrids in contact zones. In some cases such genes would be directly responsible for reproductive isolation, and in other cases they would provide the selective pressure for the completion of speciation. At least some of the structural genic variability within and between natural populations is surely maintained by natural selection. If so, structural genes themselves may contribute significantly to the adaptive divergence that often may lead to speciation.

Evolutionary change also involves regulatory genes—those genes responsible for patterns of structural gene activation and expression. Several models have been proposed that endow regulatory genes with a crucial role in evolution. As applied to speciation, it has been argued that unfavorable interactions of alleles at regulatory loci are primarily responsible for disruption of the patterns and timing of structural gene expression, resulting in decreased hybrid fitness. Regulatory gene changes could be primarily responsible for the development of reproductive isolation and hence of new species.

Allan Wilson and his colleagues famously suggested that there may be two types of molecular evolution—one involving structural genes, which goes on at a more or less constant rate; and a second involving regulatory genes, which are primarily responsible for reproductive incompatibilities and morphological evolution. This argument stems from the observation that evolutionary divergence in proteins does not closely parallel morphological divergence and the loss of hybridization potential when very different groups (such as mammals, birds, and amphibians) are compared. Although different rates of regulatory evolution are one possible explanation for such observations, it could also be that conspicuous morphological changes and reproductive isolation per se sometimes involve only a small proportion of the genome.

2

Molecular Variability and Hypothesis Testing: An Ode to Electrophoresis

The "strong inference" approach, in which competing hypotheses are formulated and critically tested, may not be the only defensible way to conduct science, but it often does help to clarify conceptual issues and refine their empirical assessments. The following paper describes the author's early attempts to wrestle with strong inference in the context of how to account for the extensive protein variation that was then being unveiled in natural populations. The laboratory technique of protein electrophoresis had been introduced to population biology in the mid-1960s, and it had revolutionized the field. For the first time, genetic variation could be assayed directly at the molecular level, in any species, without the necessity of conducting formal organismal crosses. Huge and unexpected stores of genetic variation were being uncovered in these "allozyme" surveys of numerous plant and animal taxa. Was that variation due primarily to balancing natural selection, or alternatively to genetic drift of fitness-neutral mutations?

Scientific knowledge accumulates through critical tests of hypotheses against observations gathered with the intent of falsifying those hypotheses. The base of an inductive tree consists of objectively gathered observations on which alternative explanatory hypotheses may be erected and tested. The stimulus for major advances in a scientific field often results from a novel observation or set of data, sometimes generated by the development and application of new measuring or monitoring techniques. Science proceeds most efficiently when procedures of strong inference are rigorously applied to the problem of explaining such a new set of observations.

The steps involved in conditional inductive logic, or strong inference, are as follows:

1. Alternative hypotheses are devised to explain the problem.

2. *Crucial* experiments or other empirical tests are designed with the intent of falsifying one or another of the hypotheses.

3. The experiments or tests are carried out, and false hypotheses (those inconsistent with the results of the test) are discarded.

4. Steps 1 to 3 are repeated, by generating and testing hypotheses consistent with the refined possibilities that remain.

Many conclusions in science are thus exclusions. Hypotheses are provisionally rejected whose logical implications lack congruence with the results of relevant experiments and empirical observations.

In the 1960s, for the first time, improvements in electrophoretic techniques allowed evolutionists to objectively quantify levels of genetic variability in natural populations. The results were conclusive and astounding: the genomes of individuals and the gene pools of populations are characterized by tremendous stores of genetic variability—far more than had been predicted according to some models of population genetics. This exciting discovery, and subsequent affirmation with a variety of more recent genetic analyses, generated a controversy about the causal processes responsible for maintaining all this genetic variation. At its best, the controversy contained brilliant formulations of alternative hypotheses and elegant attempts to test them. At its worst, the controversy generated uncritical data and as much heat as light. The subject of molecular variability in natural populations is of interest both for its profound implications for evolutionary processes, and as an example of the manner in which science proceeds ideally and in practice.

Basis of the Controversy: Levels of Genetic Variability

The amount of genetic variation in a population determines its evolutionary potential. The effectiveness of natural selection depends on stores of population genetic variability originally derived through mutation, recombination, and gene flow. The "fundamental theorem" of natural selection expresses the concept as follows: the rate of increase in fitness of a population at any time is equal to its genetic variance in fitness at that time. Given the signal significance of genetic variation, it may seem ironic that, until the 1960s, evolutionists could not quantify this parameter even to a first approximation.

To determine what proportion of loci in a population is polymorphic, genes must be sampled that average no more or less variable than the remainder of the genome; that is, the sample of genes must be unbiased with respect to level of variability. Yet classical Mendelian techniques determine the presence of a gene by observing segregational behavior of its different alleles in the progeny of specific organismal crosses. Genes that are monomorphic cannot be detected.

A conceptual breakthrough occurred in 1941 with the recognition that genes encode proteins. Any method that can distinguish structural properties of proteins encoded by a gene could provide secure evidence about the gene itself. One such method, protein electrophoresis, had been discovered in 1937 by Tiselius, but it wasn't until three decades later that it became a method of choice for addressing evolutionary questions. Electrophoresis separates enzymes and other proteins in an electric field, primarily according to their net electric charge. Differences in the charges of proteins are attributable to differences in the nucleotides of the codons that encode them. Proteins that differ in electrophoretic mobility are coded for either by different genes, or by different alleles if they are the product of a single gene. Importantly, proteins encoded by a particular gene and sharing an electrophoretic mobility can also be observed, and we may provisionally infer that they are encoded by identical alleles. By choosing for study several proteins encoded by independent genetic loci, estimates can be made of what proportion of genes is variable within a population, and what proportion of genes displays different alleles in populations of the same or different species.

In the early 1960s, two laboratories independently undertook the

determination of heterozygosity levels by conducting large-scale electrophoretic surveys of many randomly chosen proteins. The results
greatly surprised many people. Of 18 genetic loci in the fruit fly,
Drosophila pseudoobscura, 30% had electrophoretic variants, and an
individual was on average heterozygous at 12% of its loci. Ten genetic
loci were electrophoretically surveyed in humans, and average individual heterozygosity was 10%. Similarly high levels of genetic variability were later documented in a wide array of organisms ranging
from killer clams, horseshoe crabs, and crickets to mice and deer.

One might have thought that these exciting observations would
lead to rapid resolution of several outstanding problems in evolutionary theory, but such was not the case. To understand why, we must
first examine two schools of thought concerning levels of genetic
variability and how their proponents' views adapted to the newer observations on molecular variability.

Classical, Balance, and Neoclassical Theories of Genic Variability

The classical theory of genome structure envisions a strictly purifying
role of natural selection; selection functions to weed out the contin-
ual influx of mutations, the vast majority of which are detrimental to
the success of their bearers in leaving offspring. Of course, a few mutations may occasionally increase an organism's fitness, and these
new alleles will tend to spread through the population. Thus, evolutionary change by natural selection is by no means denied. A logical
consequence of this theory and, in fact, the cornerstone of the classical model of population structure, is that most genes in the population gene pool are fixed or monomorphic for the allele conferring
highest fitness. Most of what little genetic variation is present results
from rare adaptive mutations, recently introduced into the population and presently on their way to fixation. The classical view had re-
ceived much theoretical support, primarily from the concept of the
cost of natural selection, or genetic load.

The term *genetic load* was introduced in 1950 by H. J. Muller in an
attempt to convince the medical profession and public of the serious

health consequences of increased mutation rates. Muller was confident about the deleterious nature of most mutations and recognized that they would tend to be eliminated by natural selection. But natural selection of any form involves a cost—the price is paid in reduced population fitness and in genetic deaths. According to Muller, Kimura, Crow, and others, genetic load meant that strict limits would be imposed on the amount of genetic variability; populations necessarily must be monomorphic at nearly all loci.

The balance theory of genome structure took a different view of natural selection. Dobzhansky and his colleagues thought that natural selection frequently acted to increase and preserve genetic variability in populations. A typical population was believed to be polymorphic at a relatively large proportion of loci. Several modes of balancing natural selection were recognized as potentially capable of maintaining genetic variation: (1) heterosis, in which the heterozygote is superior in fitness to either homozygote; (2) frequency dependence, in which an allele is favored when rare but selected against when common; and (3) diversifying selection, in which selection operates in different directions in different sexes, or in different stages of the life cycle. Diversifying selection acting in different habitats may also maintain genetic variation between populations, or (when coupled with gene flow) increase genetic variation within populations. Early work on phenotypic and viability characters suggested that a great deal of genetic polymorphism was present in populations, but the evidence was inconclusive. The subsequent discoveries of high molecular variability in natural populations showed that the classical view was wrong, and triumphantly vindicated the balance view of population structure. The classical view was firmly put to rest—or so it might seem.

By an interesting shift of reasoning, however, the former followers of the classical theory regrouped and drafted a challenge that shook the balance theory to its very roots. The intellectual heirs to the classical school, the neoclassicists, did not deny that high levels of genetic variability are normally present in natural populations. Rather, they denied that natural selection plays a key role in maintaining this variability. They accepted the conclusions of the balance school but not its premises. In a sense, this challenge to the balance theory was far more serious than the former challenge of the classical school.

The neoclassicists, or neutralists, proposed that most of the newly discovered genic variability in populations is neutral with respect to fitness. In other words, it makes no difference in terms of its fitness whether an individual is homozygous or heterozygous at a given locus; the alleles are adaptively equivalent. The neutralists went further to argue that the discovery of high variability was inevitable once the use of high-resolution, sophisticated techniques such as protein electrophoresis became practical. What difference can it possible make to an organism whether it has, for example, a lysine or a glycine in position 47 of its lactate dehydrogenase molecules? Nonetheless, this difference might well be electrophoretically detectable.

According to the neutralists, most genetic variation is maintained by a balance between the influx of variability through recurrent mutation and migration, and its loss through the process of chance sampling through the generations (genetic drift). An elegant and elaborate theory was developed that predicts for any particular population the amount of genetic variability and the rates of gene frequency change through time, given the relevant parameters of mutation rate, migration rate (where applicable), and population size. Conspicuously absent from the calculations are selection coefficients (measures of fitness of different genotypes), because the alleles are assumed to be neutral. The challenge from the neutralists' prediction cut to the heart of the balance school. As Richard Lewontin wrote, "The balance school sees the maintenance of variation within populations and adaptive evolution as manifestations of the same selective forces, and therefore it regards adaptive evolution as immanent in the population variation at all times. Because the alleles that are segregating in a population are maintained in equilibrium by natural selection, they are the very alleles that will form the basis for adaptive phyletic change."

A theory or hypothesis is scientific only if there are, at least in principal, relevant observations that could falsify it. Any hypothesis that is so general or vague that it could readily explain any conceivable state of affairs does not belong within the realm of science. In other words, a hypothesis must be testable by relevant observations and experiments if it is to have scientific value or empirical content. The philosopher Karl Popper had shown that the empirical content of a hypothesis is measured by the number of its potential falsifiers. The neutrality theory made many predictions that could be tested by relevant observations and experiments. As such, the neoclassicists had made a significant contribution to science, whether or not their views were ultimately vindicated. What follows are a few examples of some predictions made by neutrality theory and how these predictions were tested by observations and experiments on allozyme variation in natural populations.

Testing Neutrality Hypotheses

Neutrality theory is able to make specific predictions based largely on mutation rates and effective population sizes. Since these parameters in particular populations are likely to be very different for different kinds of organisms, it is best to critically test neutrality predictions separately for each species or group of related species. One of the most complete early data sets was gathered from the *Drosophila willistoni* complex of fruit flies. The following arguments are based on this work, but arguments similar in principle could be cited for other organisms also. The neutralist hypotheses and their tests are presented below in only bare outline, the primary purpose being to exemplify scientific methodology in the field.

The number, n_e, of electrophoretically detectable *neutral* alleles in a population at equilibrium is predicted to be

$$N_e = (8N\mu + 1)^{0.5} \tag{1}$$

where N is the effective size of the population, and μ is the per locus per generation mutation rate to neutral alleles. Also, in a sexually reproducing random mating population, another prediction is that

$$n_e = 1/(1 - H) \tag{2}$$

where H is the average heterozygosity.

Consider the fruit fly $D.$ $willistoni$, which is distributed throughout most of Central America and northern South America. In this species, allele frequencies at polymorphic loci are often similar in different populations, even when populations are as far apart as southern Brazil and Costs Rica. Neutrality theory could explain this observation if it were assumed that sufficient migration occurs between populations to ensure allele frequency homogeneity. In that case, the entire species would behave as a single population. The size of this population is immense. In many localities, hundreds or thousands of flies may be collected with just a few sweeps of a net over banana bait. A conservative estimate of the population size for the species is 10^9.

More than 25 randomly chosen genetic loci were examined electrophoretically in $D.$ $willistoni$. Mean heterozygosity (the mean proportion of loci heterozygous per individual) is $H = 0.18$. Substituting this value into equation (2) yields $n_e = 1.2$, the observed effective number of electrophoretically detectable alleles. If these alleles are neutral, substituting the values $n_e = 1.2$ and $N = 10^9$ into equation (1) yields $\mu = 6 \times 10^{-11}$. This value for the neutral mutation rate is much lower than typical empirical estimates, which place $\mu > 10^{-7}$. The observations on $D.$ $willistoni$ do not seem in line with predictions of neutrality. To look at it differently, assume again that $N = 10^9$ (a conservative estimate) and that μ does indeed equal 10^{-7} (also conservative). Then, the predicted effective number of alleles equals 28, significantly greater than the observed n_e which is about 1.2.

Other early evidence on the pattern of allelic variation within species appeared incompatible with neutrality theory. Although, for most genes, allele frequencies are similar throughout the range of $D.$ $willistoni$, for a few genes they are very different. Neutralists predict that populations not connected by gene flow should exhibit different alleles, and in different frequencies, because of stochastic allele frequency changes through the generations. To assume homogeneity in neutral allele frequencies between populations, gene flow must be sufficient, or else not enough time has elapsed since the populations separated from a common ancestor for chance to have had a significant influence. However, genetic drift and migration affect all loci simultaneously. That is, if enough time has elapsed for some neutral loci to diverge in allele frequency, enough time presumably has elapsed for other neutral loci as well. And if migration between populations is sufficient to maintain allelic similarity at some neutral loci, all neutral loci should be roughly similar. Such was apparently not the case. Note that although all of the loci examined in *D. willistoni* could not be neutral by this train of logic, researchers could not yet say which ones are selected. In principle, it could be that genes with geographically uniform allele frequencies are under uniform natural selection, or it could be that genes with geographically heterogeneous allele frequencies experience spatially variable selection, or maybe even all genes are under one form or another of selection.

Many tests of the neutrality hypothesis similar in outline to those presented above (utilizing the observed distributions of genetic variability within and between species) were conducted. Some people were justifiably unsatisfied by such tests. They pointed out that the neutrality hypothesis could not be falsified until much more was learned about the true effective sizes of populations, mutation rates, and levels of variability across the genome. And, in any event, the results did not exclude the possibility that many loci are neutral. Other tests were thus conducted that are not subject to these criticisms, and some of them represent especially good examples of strong inference.

Natural populations of *Drosophila equinoxialis* and *D. tropicalis,* other members of the *willistoni* group of flies, are nearly monomorphic for different alleles at the gene encoding malate dehydrogenase. *D. equinoxialis* is almost monomorphic for allele "95," and *tropicalis* is nearly fixed for allele "86." Experimental cage populations were set up in which allele frequencies were artificially perturbed from their frequencies in nature. On the assumption that the alleles were neutral,

the prediction was made that the allele frequencies should remain at
their perturbed levels, because the experimental population sizes
were large and chance sampling errors would not be likely to significantly alter allele frequencies in a few generations. On the other
hand, if the fitness of an allele were a function of the physiological or
genetic background in which it acted, allele frequency changes should
be directional. After about ten generations, allele 95 significantly increased in frequency in the *equinoxialis* populations, and allele 86 significantly increased in the populations of *tropicalis*. These results
could not readily be explained by neutrality theory.

Neutralists contributed significantly to science by formulating important predictions that are eminently testable, and many valuable
studies were stimulated. In most of this research, neutrality predictions constituted the null hypotheses, that is, the hypotheses to be
tested. Predictions based on alternative modes of natural selection
have also been formulated and tested in specific evolutionary circumstances, but much more work of this sort remains before we can claim
to understand the true evolutionary significance of molecular genetic
variability.

Molecular Variation and Individual Uniqueness

Near the close of his treatise *On the Origin of Species*, Charles Darwin
wrote that through the study of evolution by natural selection, "much
light will be thrown on the origin of man and his history." Evolutionary thought has since led to the recognition of the close ties that bind
man to other biological species, and to his evolutionary past. It has
also led to the recognition of the extraordinary evolutionary uniqueness of man. The early discoveries of molecular variation also had
many profound implications for humanity's concept of itself. I will
mention just one of these.

Initial protein electrophoretic estimates based on 40 or more loci
demonstrated that roughly 40% of human genes are polymorphic.
Using a conservative estimate of 30,000 genetic loci in the genome,
this would mean that 12,000 protein-coding genes in mankind are

polymorphic. Let us assume, for the sake of argument, that, for some reason, this estimate is inflated by more than 50-fold so that only 200 loci are polymorphic. Let us further assume that each gene has only two alleles (this is extremely conservative—many loci are known to have several alleles). Rules of Mendelian heredity show that the possible number of human genotypes is then 3^{200}, or roughly a ten followed by 93 zeros. And this must be a gross underestimate!

The total number of people alive now is slightly more than 6,000,000,000. The total number of people who have ever lived is roughly 13,000,000,000. Thus, the potential number of different human genotypes vastly exceeds the number of people that have ever inhabited Earth. With the exception of identical twins, no person is likely to be genetically identical to any other living human, to any human who has lived in the past, or to any human who will ever live in the future. Studies in molecular variability have provided a biological basis for the recognition of human individuality and uniqueness.

3

The Pocket Gopher

Molecular techniques are powerful enough to identify population-genetic subdivisions within and among closely related species. This paper describes an application involving pocket gophers of the southeastern United States. In the late 1960s, the U.S. Department of Defense proposed building a Naval Submarine Base in southeastern Georgia, exactly where an endangered species of pocket gopher lived that had been described in the late 1800s. This created a problem for Georgia's Department of Natural Resources (DNR), which is charged with managing and protecting Georgia's wildlife. So, DNR officials approached John Avise and Joshua Laerm (who later became the director of the Georgia Natural History Museum) to conduct a detailed genetic assessment of this endangered species. In the course of the study, the two spent many an afternoon capturing pocket gophers under the hot Georgia sun. The process itself was interesting, as were the scientific findings that emerged. This article, written for a popular audience, explains why.

"There is a large ground rat, more than twice the size of the Norway rat. In the night time, it throws out the earth, forming little mounds, or hillocks." With these words, written in 1791, the great American naturalist William Bartram introduced us to a common rodent whose existence is closely interwoven with the soil it inhabits. The rodent is *Geomys pinetis,* the southeastern pocket gopher. The word *Geomys* is derived from the Greek roots *Geo,* meaning "earth," and *mys,* meaning "mouse." The word *pinetis* derives from the fact that the species is frequently found in stands of longleaf and slash pine, trees that are associated with the sandy soils that characterize typical habitats of pocket gophers throughout much of Florida, Georgia, and Alabama.

Bartram was not the only prominent historical figure to comment on the curious mound-building activities of these gophers. Sir Charles Lyell (whose treatise *Principles of Geology* had a profound influence on the thinking of Charles Darwin) visited the United States in 1842 and wrote: "We also saw small hillocks, such as thrown up by our (British) moles, made by a very singular animal, which they call a salamander." (The word "salamander" probably originated as a mispronunciation of "sandy-mounder," the name used earlier by plantation slaves.) To add to the confusion, "gopher" throughout the southeast is also used in reference to the burrowing tortoise, *Gopherus polyphemus*.

Most southerners know the trademark of *Geomys:* clusters of earthen mounds, each mound about 12 inches high and 1–3 feet across. These mounds are merely the tip of a burrow-system iceberg. Pocket gophers are fossorial and they spend most of their lives beneath the earth's surface. Their extensive burrows, often more than 100 feet in length, lie more or less horizontal to the surface and are from one-half to 2 feet deep (although in very soft sand they may be as deep as 6 feet). The surface mounds are the soil excavated from the burrow system. Loosened dirt accumulates in the burrow as the gopher digs. Periodically, the gopher bulldozes the soil outward to the growing mound on the surface. A single burrow system usually has six to twelve or more associated mounds spaced at intervals of several feet. The mounds are connected to the burrow system by diagonal tubes. These tubes are usually plugged with dirt by the wary gopher, so no openings are visible above ground.

The only time a burrow system is open is when a gopher makes a brief excursion to the surface to forage for grass, which the animal stuffs in external, fur-lined cheek pouches—its "pockets." Once the pockets are full, a gopher returns to the burrow where some of its food may be stored in special caches against leaner times in winter. It is not always necessary for gophers to leave their burrow systems to forage. Frequently they burrow scant inches beneath the surface collecting edible roots and bulbs. Pocket gophers in their burrows tend to be territorial and very aggressive. Despite their solitary nature, gophers do aggregate in colonies. These may include a few individuals to as many as a hundred.

Male and female gophers maintain separate burrow systems. Females tend to dig localized burrow networks, whereas males dig a more linear burrow system that may extend for hundreds of feet. Southeastern pocket gophers breed throughout the year, with peak activity in February through March and June through August. Litter size is commonly two, with a range of one to three progeny, and females may produce two litters a year. The young remain in their mother's burrow for little more than a month before dispersing. Most of their lives are spent in solitude from other pocket gophers.

Were it not for the burrowing and dietary habits of southeastern pocket gophers, they would likely have attracted little human attention. However, an active animal can do great damage to gardens, orchards, seedling trees in pine plantations, and even underground telephone and electric cables. To read the literature, one might conclude that humans had decided to wage an all-out war of extermination against pocket gophers. Between 1888 and 1976, no fewer than 203 published articles dealt with the damage incurred by pocket gophers (including western species in the genera *Geomys* and *Thomomys*). Many of these papers suggest means of controlling pocket gophers: traps, poisons, anticoagulants, repellants, gas chambers, mechanical burrow diggers, and, as one author put it, "vigil and vigor." Ironically, our interest in pocket gophers grew out of the desire to save an endangered population from extinction.

Geomys pinetis is widespread and common in the southeastern United States, but a close cousin, the colonial pocket gopher (*Geomys colonus*) appears to be near the brink of extinction. The colonial pocket gopher was originally described in 1898 by Outram Bangs, a well-known mammalogist and naturalist. At that time, the range of the species apparently encompassed scarcely a dozen square miles on the coastal plain of Camden County, Georgia; it thus was the smallest distribution of any mammal in North America. The colonial pocket gopher remained essentially unnoticed and unstudied for the next

three-quarters of a century. Beginning in 1967, Hans Neuhauser of
the Georgia Conservancy discovered mounds within the historic
range of *G. colonus,* and at least one population was delineated. This
population was estimated to be less than 100 individuals inhabiting
only 500 acres of pineland near Scotchville, Georgia.

The precarious position of *G. colonus* became exacerbated when
plans were completed to build a huge submarine base at King's Bay,
immediately adjacent to the pocket gopher colony. To officials of the
Georgia Department of Natural Resources (DNR), it was apparent
that additional information about the ecological requirements and
systematic status of the animals was needed, and quickly. Many other
pocket gopher colonies occur nearby within the traditional range of
G. pinetis, but had never been carefully studied. Could some of these
colonies in reality be colonial pocket gophers?

The job at hand was to collect and compare the DNA from many
pocket gopher populations in the area. To gain a proper perspective
on any observed genetic differences, sampling was also to encompass
most of the geographic range of *G. pinetis.* For research on the genetics of pocket gophers, we needed to capture the animals alive. This
proved to be easier said than done. An acre field may contain dozens
of mounds, some old and weathered, and it is not always clear how
many burrows are represented or whether they are presently occu-
pied. Our procedure was to dig a large circular pit around several fresh
mounds. The hole was often several feet deep and several feet in di-
ameter, and if properly excavated it would intersect the horizontal
burrow system of a gopher at two or more points in the animal's open
tunnel network. A live trap, constructed of hollow tubing equal in di-
ameter to the gopher's burrow, was then placed in each tunnel, and
within a few minutes to several hours an unsuspecting gopher might
be captured.

At first look, one might well ask whether such a capture was worth
the effort. A pocket gopher could briefly be described as a homely, bel-
ligerent sausage. Almost everything about the animal reflects adaptations for its fossorial way of life. The body is cylindrical, the eyes are

tiny, and the external ears are reduced. The turgid, nearly naked tail probably serves a dual function: It facilitates heat loss at warmer temperatures as it becomes engorged with blood; and it is actively employed during digging as a tactile organ to orient the animal and perhaps to brace the tunnel walls. The skin of *Geomys* is very loose, and the animal can seemingly swivel inside its skin to bite an owl (or human) attacker, or simply to turn in its narrow burrow.

Geomys pinetis has premolar and molar teeth for chewing roots and other parts of herbaceous material that form the bulk of its diet. More impressive aspects of the dental anatomy are the large, continuously growing incisors. The forelimbs of *Geomys* are stout and strongly clawed. They serve not only to excavate a tunnel, but also to push and pack the soil.

In studying the genetic relationships of pocket gophers, a variety of genetic techniques was employed. Chromosomes were examined under a light microscope, and the properties of numerous proteins and mitochondrial DNA were assayed by sensitive molecular techniques. All of these approaches yielded concordant results. A major genetic split, reflecting the presence of two evolutionarily distinct sets of populations, was found. One evolutionary lineage (the "eastern" form) occurs in eastern and southern Georgia and the Florida peninsula, and the other evolutionary lineage (the "western" form) inhabits western Georgia, Alabama, and part of the Florida panhandle. At the least, these two forms warrant recognition as distinct subspecies of *G. pinetis*.

The genetic differences probably accumulated during a prolonged period of geographic separation. It is quite possible that rising sea levels in the Pliocene epoch (5.3 to 1.8 million years ago) or during Pleistocene interglacial periods (1.8 to ≅ 10,000 years ago) played a role. During glacial meltbacks, rising seas are known to have inundated much of the Florida peninsula, leaving only a few islands that might have served as refuges for pocket gophers. As sea levels subsequently fell and shorelines eventually assumed their present configuration, the island populations could have spread into the present range of the eastern form. Perhaps counterpart populations on the mainland similarly expanded to assume the present geography of the western form. Numerous other vertebrate subspecies and species pairs exhibit geographic distributions similar to those of the two forms of southeastern pocket gopher. Many of these pairs currently hybridize in secondary contact zones in northern Florida and Georgia. We did not observe any hybridization between the genetic forms of southeastern pocket gophers.

In all respects, the colonial pocket gopher turned out to be clearly related to the eastern form of the southeastern gopher. In fact, *G. colonus* was indistinguishable from populations of *G. pinetis* in several surrounding counties. This cannot be due to any lack of sensitivity in the techniques employed, because each molecular method uncovered considerable genetic differences across the broader range of *G. pinetis*.

The original description of *G. colonus* in 1898 was based on morphological characters—a darker pelage and small differences in several cranial features. However, pelage color is highly variable both within and among *G. pinetis* populations, and larger collections indicate that it is not a satisfactory character for distinguishing *colonus* and *pinetis*. A total of 18 standard body measurements in more than 1,500 museum specimens of pocket gophers were compared, and *G. colonus* proved to be less distinct morphologically from adjacent populations of *G. pinetis* than are geographic populations of *G. pinetis* from one another. In the end there was no basis for recognizing the population known as *colonus* as being distinct from eastern *pinetis*. It appears to be no more than a slightly differentiated, local population of the eastern form of the southeastern pocket gopher. The logical conclusion was that colonial pocket gophers were not a separate species, and had been wrongly thought to be one for decades.

But what if this conclusion is incorrect? An ardent supporter of species status for *colonus* could always argue that critical characters separating the two species have not yet been examined. Therefore, to be safe, Georgia DNR transplanted several *colonus* gophers to undisturbed habitats where it is hoped they will continue to survive.

Looking back on this episode evokes some mixed feelings. As an evolutionary biologist, I provided honest information that allowed a natural population of pocket gophers to be disturbed. On the other hand, a great deal of money and societal effort was saved by not investing heavily in preserving a "unique and rare species" that, in fact, was neither.

4

Gene Trees and Organismal Histories

Mitochondrial DNA (mtDNA) is maternally transmitted from one generation to the next, without the interparent genetic recombination that characterizes nuclear DNA. This chapter summarizes some of Avise's early thoughts about the significance to evolutionary biology of newly discovered patterns of genetic variation in this asexually inherited molecule. Written about a decade after the author began his empirical research on mtDNA in the late 1970s, this paper reflects how the field of population genetics was then just beginning to appreciate the fundamental distinction between gene trees (genealogical histories of particular pieces of DNA) and species trees (the composite phylogenetic histories of sexual species). This realization was an important step in the phylogeographic revolution that was to help bridge the scientific gap that previously had existed between traditional microevolutionary population genetics on the one hand, and traditional macroevolutionary phylogenetics on the other. For a thorough update on this and other topics in molecular ecology and evolution, readers should consult the author's textbook, *Molecular Markers, Natural History, and Evolution*, 2nd edition (2004).

I will argue four main points in this chapter. First, that population genetics and population biology profit from the infusion of phylogenetic principles and reasoning. Second, that a powerful approach involves explicit focus on the phylogenetic histories of particular genes and gene products. Third, that an analysis of one gene system studied extensively under this philosophy—mitochondrial DNA (mtDNA)—suggests that the population sizes and the general demographic histories of most higher animals have been remarkably dynamic over recent evolution, both spatially and temporally. Finally, that an extension of this phylogenetic approach to other genes, including those in the cell nucleus, provides a satisfying conceptual framework that links the mechanistic kinds of understanding possible in molecular biology with the higher-level phenomena that are the traditional subjects of population genetics and evolution.

Population geneticists are concerned mostly with heredity and microevolutionary process; they have traditionally maintained only tenuous communication with systematists, whose primary foci typically are phylogeny reconstruction and macroevolutionary pattern. The history of and reasons for the gulf between population genetics and phylogenetic systematics are understandable, but there is irony in this state of affairs because the branches in macroevolutionary trees have a substructure that consists of smaller branches and twigs that ultimately resolve as generation-to-generation pedigrees through which genes have been transmitted. Thus, it would seem that considerations of gene history should provide a logical starting point for attempts to connect understandings of heredity and population genetics with those of phylogeny and systematics, and, thereby, to join the fields of knowledge of microevolution and macroevolution.

Some population biologists have gone so far as to argue that, since evolutionary history can never be known with certainty, it should be largely discredited in the analysis of contemporary populations. For example, Birch and Ehrlich state that "to investigate ecology and taxonomy through a series of inferences about the past is to base these sciences on non-falsifiable hypotheses," and "phylogenetic speculation(s) . . . do not help us to understand the distribution and abundance of plants and butterflies today, because they are not subject to testing." Although I agree that great care should be exercised in phylogeny reconstruction (as in any scientific enterprise), evolution is ineluctably an historical process. Genes and populations, as well as species, do have histories, and neglecting historical considerations could also lead to erroneous conclusions. For example, from a direct observation of limited contemporary dispersal in certain species of butterflies, Ehrlich and Raven conclude that "populations in Alaska are only slightly differentiated from those isolated in Colorado, indeed from those in Europe. Yet we would be greatly surprised if the Colorado populations (occurring as scattered isolates) receive a gene originating in Alaska one per hundred millennia." This may well be untrue, however, because for extended periods in the past 100,000 years, massive glaciers covered much of the northern United States and Canada. Although it is possible that some butterfly populations survived in Alaskan nunataks (unglaciated refugia within an ice sheet), it is possible that not only has gene flow occurred on a broad geographic scale, but also that entire regional populations have derived from extensive recolonizations after the retreat of the Wisconsin glaciers only some 18,000 years ago. Slatkin made clear the important distinction between contemporary and historical gene flow when he concluded from a reanalysis of Ehrlich and coworkers' genetic data on butterflies that "the current patterns are probably due to substantial gene flow in the recent past . . . due to the large-scale movement . . . permitted by unusual environmental conditions or major range expansions."

Whatever the outcome of this particular debate, it seems clear that the historical demographics of species must have left impressions on contemporary genome structure. The challenge is to use the clues of genome architecture to reconstruct historical events correctly. Traditionally, such attempts at the within-species level have been limited primarily to comparisons of allele frequencies, such as determined by protein electrophoresis. Here, I will suggest ways that a more explicit focus on intraspecific phylogeny through a complementary approach —*gene lineage analysis*—may further enrich understandings in population biology, as well as provide a continuous frame of reference for a hierarchy of evolutionary problems ranging from concern with molecular mechanisms to macroevolutionary patterns, and including adaptations.

Gene Trees Versus Organismal Trees

Awareness of the fundamental distinction between gene trees and population trees or species trees is increasing. A gene tree is the phylogeny of a particular gene or stretch of DNA. It can be estimated from nucleotide or amino acid sequences, restriction-site maps, or other procedures that view the alleles themselves as operational taxonomic units (OTUs). A population tree is the evolutionary pathway of a group of populations; populations or species are the OTUs. There are many gene trees within any population or species tree, and indeed, a population tree must in some sense represent a compilation of genealogies for many genes. Nonetheless, the estimated topology of a given gene tree can differ from that of the overall population tree because of (1) a sampling error attributable to a small number of nucleotide or amino acids examined; (2) evolutionary rate heterogeneity across gene or organismal lineages; (3) stochastic sorting of allelic lineages to daughter populations from a polymorphic ancestral population; and (4) introgressive hybridization. The last three possibilities are not simply generators of "noise" in phylogeny estimation. Rather, they are real phenomena, and a true and important part of phylogenetic history.

Various empirical issues are intertwined with these conceptual distinctions between gene trees and species trees. Until recently, allozyme products of nuclear genes provided the major source of molecular genetic data at the intraspecific level. Allozymes of a given locus are distinguished by charge differences, but the evolutionary relationships (gene trees) of the alleles thus identified cannot be safely inferred from the observable property: electrophoretic mobility. Thus, it has been customary and, in most cases, necessary to restrict attention to allozyme frequencies, which represent composite attributes of assemblages of individuals. Composite attributes of assemblages of genes are reflected in estimates of genetic divergence derived from certain other molecular approaches, such as multilocus protein electrophoresis or DNA/DNA hybridization; but again, information about particular gene phylogenies is missing. On the other hand, molecular assays such as amino acid sequencing or nucleotide sequencing, which do provide information relevant to construction of allelic genealogies, have previously been applied almost exclusively to the estimation of relationships among higher taxa, where the issue of intraspecific polymorphism can be safely neglected.

With the advent of mtDNA assay techniques, it became feasible to estimate gene trees even at the intraspecific level. The relevance of microevolutionary allelic trees (both mitochondrial and nuclear) to population biology is the subject of this chapter.

Mitochondrial DNA and Intraspecific Gene Phylogeny

Two major factors explain why mtDNA has been by far the most extensively exploited molecule for microevolutionary gene-lineage analysis in higher animals: it evolves very rapidly, primarily through base substitutions; and it is cytoplasmically housed, maternally inherited, and effectively haploid in transmission across generations. These latter attributes mean that individual organisms (as well as mtDNA alleles or haplotypes) can justifiably be considered as the OTUs in a phylogeny reconstruction, which is then interpreted as an estimate of a matriarchal tree. In terms of function, mtDNA is composed of 37–38 genes or coding regions; but from a phylogenetic perspective, because mtDNA is nonrecombining, the entire 16- to 20-kilobase molecule represents one genealogical unit.

Within several assayed vertebrate groups, this genetic unit is reported to evolve at a mean rate of roughly one percent sequence change per lineage per million years. Beyond about 5–10 million years, however, the overall rate at which related mtDNA lines accumulate differences from one another gradually declines, as the faster-evolving positions in the genome become saturated with nucleotide substitutions. Base changes accumulate most rapidly in portions of the control region and most slowly in the tRNA and rRNA genes. Whether the mtDNA rate calibrations and patterns cited above apply to all vertebrate groups and to invertebrates are topics still under debate. In any event, it is abundantly clear that most species of higher animals harbor a wealth of intraspecific mtDNA polymorphism that can be tapped (e.g., by sequencing) to reveal relationships among mtDNA alleles or haplotypes.

A mitochondrial DNA phylogeny consists of the proposed historical relationships among haplotypes deduced from the numbers and sequences of mutational changes in the molecule. Numerous algorithms exist for constructing trees or networks from mtDNA databases, but these various alternatives will not be addressed here.

Two major aspects of intraspecific mtDNA variability are of interest: the magnitude and pattern of phylogenetic differentiation among the mtDNA haplotypes themselves, and the geographic distributions of the mtDNA phylogenetic groupings or clades. Together, these aspects constitute concerns of a broader discipline that my colleagues and I have termed intraspecific phylogeography. In principle, the magnitude of mtDNA phylogenetic divergence and the degree of spatial structure are independent variables among species, such that they can be plotted on orthogonal axes. Thus, for a given species, intraspecific nucleotide-sequence divergence between mtDNA haplotypes could be great, and the mtDNA clades could either be strongly partitioned geographically (phylogeographic category I) or weakly partitioned geographically (category II, in which mtDNA clades are geographically widespread and in similar frequencies among populations). Conversely, sequence divergence between haplotypes could be small, and the mtDNA clades could either be strongly or weakly structured in geographic space (phylogeographic categories III and IV, respectively).

Empirical data from intraspecific mtDNA phylogeography provide valuable insights. Many assayed species fall into phylogeographic category I; mean mtDNA sequence divergence between mtDNA clades is great (e.g., more than about 2%), and the clades are geographically distinct such that regional populations of a species belong to separate main trunks of the intraspecific mtDNA phylogeny. For example, there are two main branches in the mtDNA network for the spotted sunfish (*Lepomis punctatus*) that differed by at least nine assayed mtDNA mutation steps and an estimated 4.4% sequence divergence (after correction for within-region clonal divergence). Representatives of one branch of the phylogeny appeared to be confined to drainages east of the Apalachicola River in the southeastern United States, while members of the other phylogenetic branch were observed in all assayed fish from the Apalachicola west to Louisiana. Other species fall into phylogeographic category III, where mean mtDNA sequence divergence is relatively low (e.g., less than about 1%), yet the mtDNA clades (which were more weakly defined) remained geographically separate. A good example involves the diamondback terrapin (*Malaclemys terrapin*), in which two mtDNA haplotypes, differing by a single assayed restriction site and approximately 0.2% in assayed sequence, characterize populations in salt marshes north versus south of a region in northeastern Florida. Some species exhibiting phylogeographic pattern IV have also been observed. There, sequence differences between mtDNA genotypes and clades are small and the geographic populations show little or no evidence of major differences in haplotype frequency. The best available example involves the American eel (*Anguilla rostrata*), a catadromous species whose peculiar life history has the effect of providing extensive gene flow between geographic areas.

Because the precise dividing lines between the phylogeographic categories are arbitrary, some species may be expected to fall into gray areas in the classification scheme of population structure. For example, humans on a global scale, and red-winged blackbirds (*Agelaius phoeniceus*) on a continental scale, both exhibit low mtDNA sequence divergence and "some" geographic localization of haplotypes (i.e., significant but not fixed haplotype frequency differences distinguishing some regional populations). Of special interest is the fact that very few species have been reported to show phylogeographic category II: high intraspecific mtDNA sequence divergence in the absence of geographic localization of the mtDNA clones or clades.

To interpret such phylogeographic patterns in terms of the demographic histories of species, various kinds of probability models addressing allelic relationships within and between populations are relevant. Consider, for example, the expectation for the mean probability distribution of times to identity-by-descent for pairs of selectively neutral mitochondrial DNA haplotypes within idealized random-mating populations, or within entire species characterized by very high gene flow. This can be obtained from a slight modification of conventional inbreeding theory. Suppose that in each generation G, females contribute to a large gamete pool from which a random draw of size N produces daughters. The probability that a randomly chosen pair of such offspring then shares a common mother is $1/N$, which is also the probability (F) that mtDNA haplotypes are identical by descent from the prior generation. Mitochondrial DNA alleles could also be identical by descent through replication in earlier generations. Thus, F accumulates across time, eventually reaches 1.0, and in any generation is given by

$$F_G = 1/N + (1 - 1/N)\, F_{G-1} \qquad (1)$$

The probability distribution of times to common ancestry, which can be generated from equation (1), is geometric with mean N and variance $N\,(N-1)$. In such idealized populations, it is assumed that constant numbers of breeding females produce random numbers of daughters and are replaced in each generation, such that the distribution of family sizes is approximated by the Poisson distribution with mean 1.0. In practice, for comparisons with real populations in which these assumptions are unlikely to hold exactly, N in equation (1) can be replaced by N_e, the evolutionary effective population size of females.

Times to common ancestors of different pairs of haplotypes from a single gene (such as mtDNA) are correlated due to coancestry in the population pedigree. Hence, the theory outlined above, which assumes genealogical independence among haplotypes, does not strictly apply for particular gene genealogies within an organismal pedigree. Nonetheless, these theoretical expectations do hold reasonably well, and they can be used to generate expectations for haplotypes relationships to a first approximation.

For each of three currently abundant vertebrate species (American eels, hardhead catfish, and red-winged blackbirds) with historically high gene flow, such that the entire species is the relevant unit of analysis, effective population size (N_e) was estimated from conventional mtDNA clock calibrations (see above) as applied under inbreeding theory to observed distributions of nucleotide-sequence differences among mtDNA haplotypes. In all cases, present-day census sizes (N) proved to be much larger (by 100- to 1,000-fold) than inferred N_e values, indicating that mtDNA genotypes were channeled through far fewer ancestors than might otherwise have been anticipated.

In fact, for any abundant species with high gene flow, the theoretical expectation is that many mtDNA haplotypes should be highly divergent genetically, provided evolutionary effective population sizes have also been large. For example, in an annual species with N_e = 5,000,000 (not an unreasonable guess for the current breeding population size of eels, catfish, or red-winged blackbirds), the mean time of separation of randomly drawn mtDNA alleles should be about 5,000,000 years, which translates under the conventional mtDNA clock calibrations to p = 0.10 base substitutions per nucleotide. However, as already mentioned, there are very few examples in the literature (apart from secondary hybrid zones) of panmictic populations or of quasipanmictic species exhibiting large mtDNA phylogenetic gaps (phylogenetic category II). Thus, if we can generalize from available mtDNA data for species with large population sizes and high rates of gene exchange, N_e commonly appears to be vastly smaller than N. Possible explanations (not mutually exclusive) for this result include: high variances in progeny survival among females; intense selection resulting in the fixation of one or more advantageous mutants in the recent past; dramatic fluctuations in population size, perhaps including very recent expansions to current levels; or other related demographic factors that had the effect of depressing the numbers of female ancestors that contributed to the current gene pool through female lines.

Another class of useful probability models in the analysis of gene lineage relationships involves application of generating functions to distributions of family size in branching-process theory. If, for example, females again produce daughters according to a Poisson distribution with mean μ = 1.0, the probability of random loss of any female line in one generation is e^{-1} = 0.37, or in general, e^{-u}. The probability of loss of a line after G generations is given by the generating function $P_G = e^{u(x-1)}$, where x is the probability of loss in the prior generation. Generating functions have been used to examine the expected fates of neutral mtDNA alleles within populations as a function of demographic conditions such as population size and offspring distribution; results showed that random lineage extinction is a surprisingly important and usually rapid process. For example, for a stable-sized population initiated by N females or density-regulated at a carrying capacity $K = N$, after approximately N generations the probability is nearly 50% that all mtDNA alleles will trace back to one founder haplotype. Such alleles would then have a monophyletic origin within that time span. Generating functions are also available for family-size distributions other than the Poisson, such as the negative binomial. A general result is that, for any mean number of progeny, as the variance in progeny survival across families increases, the rate of mtDNA lineage extinction also rises, and the time to population monophyly decreases.

The extinction of mtDNA (or other allelic) lineages under branching-process theory is a stochastic but nonetheless inevitable consequence of population turnover through reproduction. Gene trees within a population are continually "self-pruning," with the net effect of continually truncating the frequency distribution of times to common mtDNA ancestry. The limited mtDNA sequence divergence observed within most local populations, or within entire species characterized by high gene flow, is likely attributable in large part to this process.

However, many species probably consist of geographic populations behaving as quasi-independent or completely independent demographic units because of severe historical restrictions on gene flow. For any such species considered as a whole, the preceding theory for individual populations does not apply directly and must be modified to take into account the fact that the times of separation of mtDNA lineages among completely isolated populations could be no less than the times since separation of the populations themselves. Furthermore, for any two populations that have separated recently, it is quite likely that some individuals in population A are more closely related in the mtDNA haplotype phylogeny to certain members of population B than they are to other members of A, due solely to stochastic sorting from the ancestral pool of lineages.

By using computer simulations in an extension of the branching-process theory outlined above, probabilities of various phylogenetic relationships (reciprocal monophyly, paraphyly, and polyphyly) have been examined for mtDNA lineages in two isolated populations descended from common ancestral stock. Outcomes highly depended on the particular demographic conditions specified, but, in general, populations tended to evolve to a phylogenetic status of reciprocal monophyly within, at most, about $4N$ generations of isolation. At separation times less than that, populations were commonly paraphyletic or polyphyletic with respect to the mtDNA haplotypes lineages segregating within them.

Empirical data have demonstrated that local and regional populations within many assayed species occupy separate identifiable twigs and branches, respectively, in an intraspecific mtDNA phylogenetic tree. According to theory, such populations have likely been isolated from one another for more than N_e generations. For sedentary species with limited gene flow (such as the pocket gopher, *Geomys pinetis*), genetic drift and population bottlenecking at a very local level probably yield quite small effective population sizes, such that mtDNA lineages become quickly but transiently fixed on a fine geographic scale. An example is the geographic clustering and localization expected for individual family groups, some of which are detectable in conventional mtDNA assays. Over longer periods, however, many of these twigs of the tree are lost while others increase in frequency and geographic range, such that the kaleidoscope of local genetic differences is continually shifting.

On a broader geographic scale and time, local populations that were historically connected by gene flow group into more extended branches of the mtDNA phylogenetic tree. Not infrequently, populations over large geographic regions occupy phylogenetic branches highly distinct from other such regional assemblages. The separation of such major branches probably evidences long-term zoogeographic barriers to gene flow, allowing accumulation of haplotype differences greatly exceeding those observed within local populations or regions. Species in phylogeographic category I typically have between-region mtDNA differences greatly exceeding mean mtDNA distances within local populations or even within entire panmictic species. In spotted sunfish, for example, the mean mtDNA sequence divergence between phylogeographic assemblages (p = 0.062) was considerably greater than the mean (p = 0.018) or even maximum (p_m = 0.042) genetic distance observed within a geographic clade. Using mean within-region genetic distance as a correction factor for the between-region differences, the mtDNA sequence divergence estimated to postdate regional population separation in *L. punctatus* becomes p = 0.044, which according to the conventional mtDNA "clock" suggests that these populations diverged more than 2,000,000 years ago.

Gene Phylogeny and Historical Population Demography

The overall picture to emerge from the numerous mtDNA studies of intraspecific phylogeography is one of dramatic dynamism and flux of mtDNA lineages. Many species exhibit well-defined local structure with respect to mtDNA genotypes, suggesting that there are restrictions on gene flow and relatively small effective population sizes of matrilines. On a regional geographic scale, the common presence of major mtDNA phylogenetic gaps indicates the profound effects of zoogeographic barriers in shaping extended pedigrees. Even for those species that are currently large and panmictic or quasipanmictic, the time-depths of mtDNA lineage separations suggest evolutionary N_e values that are vastly smaller than current N values, perhaps because such species have fluctuated greatly in population size in the recent evolutionary past. Taken altogether, the results of the mtDNA lineage analyses imply that population demographies of many species have been dynamic over time and space.

Is this picture of historical demography correct? Admittedly, for any of the species assayed, particular inferences about the demographic past were necessarily based on patterns of mtDNA lineage relationships in extant populations. However, it is a simple matter to cite contemporary, directly observed examples of the qualitative kinds of irregular demographic events that we propose have occurred in the evolutionary histories of many if not most species. The following are examples from my own experience, or are well-documented cases in the literature; I have included them merely to convey an impression of some of the extreme sorts of demographic occurrences that may profoundly influence the structure of intraspecific gene phylogenies at both local and regional geographic scales.

Extinction or near-extinction of local populations is probably a common occurrence in many species because of changes in climate and physical habitat or biotic factors such as competition and disease. This point was forcefully driven home to me in August 1980, when during a detailed study of microgeographic clonal diversity and recruitment in staghorn coral (*Acropra cervicornis*) at Discovery Bay, Jamaica, the entire population (and associated forereef environment) was devastated by Hurricane Allen. Other local *Acropora* populations are also known to have been decimated by hurricanes and cold fronts. At any given reef locale, such major physical disturbances may well occur, on average, once every 100 or 1,000 years, well within timescales relevant to influencing intraspecific genetic architectures.

Such extinctions need not be confined to small areas. For example, prior to 1983, the black sea urchin (*Diadema antillarum*) was one of the most abundant invertebrates on reefs throughout the Caribbean, sometimes reaching densities of 70 individuals per square meter. Within one year, mass mortality, apparently due to a virulent and species-specific waterborne pathogen, reduced population densities throughout the 3.5 million-square-kilometer area of the Caribbean to about one percent of former levels. Devastating disease outbreaks of this sort probably afflict a great many species over their evolutionary life spans. Palumbi and Wilson have recently surveyed mtDNA variability in other sea urchin species (genus *Strongylocentrotus*) and, from an analysis similar to those discussed earlier in this chapter for eels, catfish, and red-winged blackbirds, concluded that the observed sequence diversity was "too low to be explained by a simple application of the neutral theory of genetic change, [suggesting] either that mtDNA variants have been under strong selection in the recent past, or that sea urchin population sizes have undergone dramatic fluctuations." Whether genotype-specific selection per se or population-size variation is the responsible factor in any of these instances will be difficult to determine, because both processes would have the net effect of channeling mtDNA lineages through fewer females and hence of reducing apparent N_e.

In any event, several other cases of mass mortality of marine species are known. For example, climatic changes associated with El Nino have had dramatic effects on many Pacific species. Since physical environments in lakes and on land are usually even more variable than those in the ocean, freshwater and terrestrial species are probably at least as susceptible to dramatic population size reductions.

Conversely, dramatic population and range expansions are also certainly frequent enough, relative to the mean evolutionary life spans of many species, to play a significant role in shaping the intraspecific structures of gene lineages. The comparatively minor mtDNA haplotype divergences among humans on a global scale and among red-winged blackbirds across North America may be examples of the expected consequences of extensive population expansions and movements in late Pleistocene and Recent times. The effects of such range expansions may of course also be exhibited at other geographic scales. For example, my coworkers and I proposed that the regional uniformity of mtDNA haplotypes in *L. punctatus* and in other fish species of the southeastern United States, relative to the magnitude of between-region haplotype differences in each species, reflects zoogeographically constrained dispersal out of disjunct Pliocene or Pleistocene refugia.

Again, these particular inferences are based on present-day distributions of mtDNA lineages, but innumerable examples of directly observed range expansions in other species could be cited. For example, the cattle egret (*Bubulcus ibis*) is now the most plentiful egret in North America, but the continent was colonized only 30 years ago by as few as 14 immigrants to Florida, who themselves were descendents from nineteenth century colonists of South or Central America from native African stock. Many such range expansions in birds are well documented, such as the dramatic spread since 1942 of house finches (*Carpodacus mexicanus*) in eastern North America, and the occupations of North America within the last century by the house sparrow (*Passer domesticus*) and European starling (*Sturnus vulgaris*). Birds typically have high dispersal capabilities, but even species with much more limited mobility unquestionably experience major shifts in geographic range through time. The past two million years, in particular, have been a time of exceptional climatic instability, as the direct and indirect effects of glacial advances and retreats altered species' distributions in low and high latitudes.

The point of this argumentation is that much evidence points to the dominant role of historical demographic and zoogeographic factors in shaping intraspecific gene phylogenies. Direct observations of contemporary populations provide numerous examples of impressive fluctuations in population size and geographic distribution; climatic and geologic changes over recent evolutionary time have undeniably had a major impact on species' abundance and distribution; and available data on phylogeographic patterns of mtDNA genotypes in many species are consistent with interpretations of historical demographic influence. Continued explorations of intraspecific gene phylogenies (in conjunction with the more traditional study of population trees through allele-frequency analyses) offer a realistic hope for reconstructing the historical events that have necessarily influenced genetic structure in contemporary populations of any species.

Summary

In any scientific discipline, one must decide what constitutes an adequate and satisfying level of understanding of natural phenomena. It is probably fair to say that systematists have set forth a broad goal of elucidating the phylogenetic and evolutionary histories of taxa; population geneticists are concerned with the processes responsible for gene-frequency changes within and among populations; and molecular geneticists are concerned with gene structure and function. A comprehensive understanding of evolution must include input from all of these (and other) perspectives. My aim has been to suggest ways that a relatively new kind of population data (on DNA haplotypes) may provide empirical and conceptual connections between several disciplines. An expanded concern with this gene-phylogeny approach may accomplish several objectives: enrich population biology and population genetics by adding the dimension of intraspecific evolutionary history (which has too often been neglected) to interpretations of contemporary population genetic structure; provide feedback to molecular genetics by allowing deductions about molecular-level phenomena (such as recombination, or the molecular basis of particular adaptations); and expand the sphere of systematics to include microevolutionary genealogy and the intraspecific pedigrees that constitute the phylogenetic substructure of evolution.

5

Nature's Family Archives

Since the late 1970s, mitochondrial DNA (mtDNA) has been an important tool for genealogical research in evolutionary biology. Key features of the molecule in most animals are its rapid evolution and its asexual mode of transmission through female lines. Indeed, some remarkable genealogical parallels exist between maternally inherited mitochondrial lineages and paternally inherited human family surnames. This chapter comes from a paper written for a general audience in which Avise draws explicit connections between these seemingly different kinds of historical information.

Family names were first used in China during the Han Dynasty (about the time of Christ), but the widespread practice of assigning hereditary surnames to family lines came relatively recently to most parts of the world. In England, surnames were not customary until at least the fourteenth century, and in Japan, only the governing classes were allowed surnames until 1875, when a cabinet decree mandated their adoption by the entire populace. Hereditary surnames help to organize family records in complex societies. They also provide a chronicle of recent human history—a record that can be read to reveal patterns of human dispersal and settlement.

Evolutionary biologists have a similar record for other animal species, and can trace the lineages of a species back through time to reveal the origin and subsequent dispersal of family groups. This record resides in especially clear form in the molecular makeup of a piece of genetic material called mitochondrial DNA (or mtDNA).

Although both sons and daughters take the surname of their father, only sons pass it on to the next generation in most families. Analogously, sons and daughters both inherit the mtDNA of their mother, but only daughters pass mtDNA to their progeny. This is because mitochondria—cell components sometimes called the "power plants of a cell"—are found not in the cellular nucleus, but outside it in the cell's cytoplasm, away from the chromosomes. When sperm and egg unite to form a new organism, they contribute equally to the genetic material in the new cell's nucleus, but for the most part, the cytoplasm is the contribution of the egg alone. Males are therefore as irrelevant to mitochondrial heredity as females are to traditional human surname heredity. This single parent transmission greatly simplifies evolutionary bookkeeping.

As mtDNA makes copies of itself, sometimes independently of cell division, mutations occur. By counting the number of mutations that distinguish populations of known age, one can determine the approximate rate at which mtDNA differences accumulate. With the rate of evolution calibrated, one can then estimate the length of time that any other lineages have been separated from one another by comparing their mutational differences. In vertebrates, mtDNA typically evolves up to ten times faster than nuclear DNA. Many mtDNA mutations appear neither to harm nor help their carriers, but much like surnames—Smith, Smyth, Smithers, for example—they simply distinguish one group from another. Most species have a great number of these idiosyncratic genotypes, which can give clues to the place of a species' origin just as a surname may indicate nationality.

For example, based on comparisons of mtDNA among desert tortoises of the North American southwest, biologist Trip Lamb and I found evidence for three population subdivisions. One group lives in southern Sonora, Mexico; the second lives in desert and subtropical scrublands throughout western and central Arizona and northern Sonora; and the third apparently lives only in southern California, Nevada, and Utah. The Colorado River valley separates the last two. The "clock" calibrations suggest that the mtDNAs of these three groups have been separated from one another for several million years.

The timing of the split between two major branches in the tortoise's mtDNA phylogenetic tree can be related to the geology of the

Colorado River basin. About six million years ago, a 20-to 30-mile-wide embayment of brackish waters, the Bouse Sea, reached northward from the Gulf of California to the current Lake Mohave area in southern Nevada. The sea may have split the ancestral population of the desert tortoise. Subsequent geological uplifting resulted in the sea's retreat and the formation of the modern Colorado River, which, until dammed in this century, probably continued to separate the tortoise populations.

Freshwater fishes in the southeastern United States exhibit similar genetic breaks. Eldredge Bermingham, while a graduate student in my laboratory, surveyed the spotted sunfish in rivers along the coastal plain from the Carolinas to Texas. Analysis of mtDNA revealed seventeen different "family names" that could be grouped into two matriarchal lines. Spotted sunfish from west of the Apalachicola River (part of the state boundary between Alabama and Georgia) were genetically distinct from those in South Carolina, Georgia, and the Florida peninsula. Four other species of fish—the bowfin, warmouth sunfish, bluegill sunfish, and redear sunfish—showed similar geographic splits.

Within the past few million years, several glacial periods froze vast quantities of water at higher latitudes. Sea levels dropped and exposed regions of the southeastern coast of the United States that today are under the ocean. At such times, certain rivers that are now separate may have converged at their lower reaches (much as the Alabama and Tombigbee rivers now meet just before they flow into the Gulf of Mexico at Mobile, Alabama). At such times freshwater fishes swam from one river to the other and populations mingled. During the interglacial periods, sea levels rose and again inundated the coastal plains, and fish sought refuge upstream, creating isolated groups within each species. When the sea level went down, certain groups again dispersed and combined. Through it all, the Apalachicola region, which is higher than the surrounding land, appears to have been a relatively insurmountable barrier. Fish to the east and west of it may be the same species, but they represent different genetic groups.

And over time, with divergence in their nuclear genes as well as in their mtDNA, their physical characteristics have changed so that populations are often described as separate subspecies.

Sometimes, evidence of relatedness gained through analysis of mtDNA contradicts previous assumptions about an animal's evolutionary history. Mammalogists had long tried to identify the various subspecies of deer mouse, one of the most ubiquitous small mammals in North America. They based their classifications on such traits as fur color and ear and tail lengths, characteristics presumably influenced by nuclear genes. When Bob Lansman and I analyzed the mtDNA of deer mice, we found 61 different genotypes among 135 animals assayed from southern Mexico to the Northwest Territories of Canada, and from the Pacific to the Atlantic seaboards. Some of the matriarchal "surnames" were much more closely related than others, and these proved to have distinct geographic distributions. One group of related mice was confined to the Appalachians and adjacent locales in the eastern United States; a second was found throughout the Central Plains, the Rockies, the Pacific Northwest, and western Canada; a third was observed only in southern California. These mtDNA lineages often bore little correspondence to the already named subspecies.

The reason for this lack of correlation, I believe, is that certain physical characteristics, such as coat color, often represent adaptations to local conditions. Such traits can arise independently in distantly related (or even unrelated) lineages and are therefore poor guides to genealogical history. By contrast, because mtDNA differences involve such a large number of genetic characters, the odds are very small that so many mutations could arise in the same way twice. It is more likely that any widespread similarities in mtDNA are due to shared matriarchal ancestry.

Animals with similar mtDNA genotypes are not always confined to the same geographic area. For example, different mtDNA genotypes are distributed widely among populations of American eels between Maine and Louisiana. The eel is catadromous, that is, it spends most of its life in freshwater and travels to the sea to spawn (the reverse of anadromous species such as salmon). At sexual maturity, eels migrate to an area in the western mid-Atlantic Ocean known as the Sargasso Sea. Here, they spawn en masse. The larvae, transported by ocean currents, then drift for about a year, returning to American coastal waters. The mtDNA data show that eels collected from various freshwater rivers in North America represent random genetic draws from a single gene pool for all American eels.

Eels from Europe also migrate to the mid-Atlantic to spawn, and because they are physically nearly indistinguishable from American eels, many biologists thought they interbred. But the genetic assays show that American and European eels are largely separate breeding populations, perhaps because they spawn in separate areas of the Sargasso Sea.

Besides allowing mtDNA histories to be studied in any animal species, in other respects, too, the evidence of biological ancestry provided by mtDNA is superior to that provided by surnames. First, adopted or out-of-wedlock sons who assume the family name of their mother's legal husband introduce errors into the surname record of a human genetic pedigree. Second, human surnames often have multiple independent origins in different places—Baker has arisen in many unrelated families because the source of that surname involves a common profession. Likewise, last names can be derived from a prominent historic figure (like Williams), or conversion from earlier systems of patronymics in which a name was formed by the addition of a suffix to the given name of one's father (like Johnson). In contrast, each distinctive mtDNA lineage has arisen only once in evolution.

Since mtDNA evolves so rapidly, what controls the number of mtDNA types within a species? The answer is that since the organism must reproduce to pass on genes, matriarchal lineages also continually go extinct. The mtDNA of any female who fails to leave female descendants through daughters will forever be lost. Again, an analogy from human surnames may help to clarify this concept. Pitcairn Island in the Pacific was colonized in 1790 by six male mutineers of the

British ship *Bounty* and thirteen Tahitian women. Few outsiders have ever gone to Pitcairn to live. A recent population of fifty individuals was only six or seven generations removed from the founders, yet surnames of only three of the mutineers remained (plus the surname of a whaler who later settled there). Thus, 50% of the founding surnames had already gone extinct, and in time, if the island remains completely isolated, a single surname may take over entirely.

Such a model of lineage survival and extinction is identical for mtDNA. Lineage extinction can be rapid. Suppose, for example, that a stable-sized population is founded by 100 females, each with a distinct mtDNA genotype, and that in every generation mothers produce daughters according to a reasonable expectation. In the first generation, approximately 37 mtDNA genotypes will by chance go extinct (mothers who happen to leave no daughters), and within 20 generations, only about 10 mtDNA genotypes will remain. Thus, mtDNA trees within a species are continually "self-pruning." The mtDNA diversity within a species is a compromise between the rate of origin of new types through mutation and the rate of extinction of existing types.

mtDNA and surnames are wonderfully simplified systems of information storage because they follow strict matrilineal and patrilineal inheritance, respectively. But such particular genealogical tracings represent only a small fraction of the total hereditary history of a species. So, in the future, similar molecular analyses should also be extended to particular genes in the cell nucleus.

6

Molecular Clones within Organismal Clones

Evolution has produced a rather large number of vertebrate "species" that consist solely of females. For several years, Avise joined forces with several other researchers to study the genetic origins and evolutionary histories of these unisexual taxa, which typically reproduce by asexual mechanisms such as parthenogenesis. These remarkable organisms are of interest in their own right, and they also provide wonderful fodder for addressing the evolutionary ramifications of nonrecombining biological systems. The article on which this chapter is based was a review of genetic studies on asexual mitochondrial lineages within asexual organismal lineages (i.e., of clones within clones). In sharp contrast to the situation in sexually reproducing taxa, the transmission history of mitochondrial DNA (mtDNA) (as well as all nuclear loci) in any asexual species is in principle one-and-the-same as an entire organismal genealogy, and this peculiar fact offers unique advantages for deciphering the evolutionary geneses and properties of unisexual taxa.

Some vertebrate "species" exist predominantly or exclusively as females, exhibiting asexual or semisexual reproduction. Examples occur among the fishes, amphibians, and squamate reptiles. Essentially all known unisexual vertebrates carry the nuclear genomes of two or more bisexual species, and thus arose via interspecific hybridization. These all-female "biotypes" reproduce without genetic recombination, by one of three modes: (1) parthenogenesis, in which the female's nuclear genome is transmitted intact to the egg, which then develops into an offspring genetically identical with the mother; (2) gynogenesis, in which the process is the same except that sperm from a related bisexual species is required to stimulate egg development; and (3) hybridogenesis, in which an ancestral genome from the maternal line is transmitted to the egg without recombination, while paternally derived chromosomes are discarded only to be replaced in each generation through fertilization by sperm from a related sexual species. For the sake of clarity, a note about terminology is required. The usual definitions of biological species do not easily apply to unisexual forms. "Biotype" will be used in this chapter, although its meaning also may be unclear unless the genomic constitution of a hybrid unisexual form is specified. For the sake of continuity with the literature, I will employ traditional Latin binomials where they have been assigned to unisexual biotypes, and use hybrid genomic designations where they have been applied.

Conventional wisdom holds that the rarity of unisexual reproduction in higher animals stems from both proximate and evolutionary factors. The window of opportunity for production of unisexual biotypes may be quite narrow—presumably, genetic differences between the hybridizing taxa must be sufficient to disrupt recombinant processes during gametogenesis, but not so great as to severely impair viability, fecundity, or other fitness components. Longer-term evolutionary constraints presumably involve a paucity of genetic variation by which unisexuals might adapt to changing environments, and the accumulation of deleterious mutations and gene combinations that cannot be purged in the absence of genetic recombination. Nonetheless, approximately 70 unisexual vertebrate biotypes have been identified, and they are common in some groups, such as *Cnemidophorus* lizards. Some unisexuals also have large populations and occupy extensive ranges, suggesting that asexuality can be a successful evolutionary strategy at least in the short term. How frequently do unisexual biotypes arise? What are the mechanics of their origin? Which bisexual species provided the male and female parents in the original hybridizations? How long do unisexual lineages survive? Answers to these and related questions are of interest in their own right, and may offer broader insights into the significance of sexual reproduction.

The mitochondrial DNA (mtDNA) molecule is favorable for analyses of unisexual complexes because it evolves rapidly in nucleotide sequence and exhibits maternal, nonrecombining transmission through organismal pedigrees. Thus, mtDNA provides a common genetic yardstick by which to compare the magnitudes and patterns of maternal lineage separation in populations both of unisexual biotypes and their sexual progenitors. Furthermore, unlike the situation in sexually reproducing species, where each gene genealogy represents only a minuscule component of the organismal phylogeny, the transmission pathway of mtDNA within a unisexual biotype is in principle one-and-the-same as the organismal pedigree.

To unravel the unisexual complexities mentioned above, genetic analyses of mtDNA have been conducted on more than 25 unisexual vertebrate biotypes and their sexual relatives. Typically, sequence divergence values between mtDNA haplotypes were calculated and estimates of phylogeny were generated from these genetic distances as well as from the raw data.

Is mtDNA Strictly Maternally Inherited?

A unisexual fish, *Poeciliopsis monacha-lucida,* provided an unusually critical test of the possibility of paternal leakage of mtDNA in a hybridogenetic system. In the laboratory, crosses of *P. monacha* females × *P. lucida* males sometimes produce viable hybridogenetic lineages, whereas the reciprocal matings do not. Naturally occurring strains of all-female *P. monacha-lucida* presumably arose in this same fashion, and are perpetuated by hybridogenetic reproduction involving sperm from *P. lucida* males. Thus, in effect, the unisexual biotype is maintained by continued backcrosses to *P. lucida,* although the *P. lucida* nuclear genome is lost prior to each hybrid meiosis. If even a minute fraction of zygote mtDNA derives from sperm in each generation, natural hybridogenetic strains should have accumulated considerable *P. lucida* mtDNA over evolutionary time. Nonetheless, assays conducted in our laboratory consistently have failed to detect *P. lucida* mtDNA within natural or synthetic strains of *P. monacha-lucida*. Results are consistent with strict maternal inheritance of mtDNA in these fishes.

Unisexual vertebrates that reproduce by gynogenesis are probably less susceptible to sperm-mediated mtDNA input, since fusion of egg and sperm nuclei does not take place. In the laboratory assays, we have found no evidence for paternal leakage of mtDNA in gynogenetic triploid forms of *Poeciliopsis*. The role of sperm during fertilization of these fish is unknown. Does it just contribute a mechanical signal to divide, or does it provide an essential chemical input? Parthenogens do not require sperm at all, and thus are not susceptible to paternal leakage of mtDNA. So, for the remainder of this chapter, we assume strict maternal inheritance of mtDNA for all unisexual taxa considered.

Direction of Hybridization Events Producing Unisexuals

Although the bisexual progenitors of most unisexual vertebrates were known or suspected from earlier comparisons of morphology, karyotype, allozymes, geographic range, or other information, the direction(s) of crosses had remained conjectural in essentially all cases. Mitochondrial assays have closed this information gap. For example, *Poecilia latipinna* and *P. mexicana*, which are the known sexual progenitors of the gynogenetic fish *P. formosa*, proved to differ substantially in mtDNA, whereas the mtDNA of *P. formosa* is essentially indistinguishable from that in *P. mexicana*. Thus, *P. mexicana* was the female parent of the assayed gynogens.

Similar mtDNA inspections by various researchers have allowed unambiguous determination of the female progenitor for more than 25 unisexual biotypes. Typically, the bisexual relatives of the unisexuals are highly distinct in mtDNA genotype, whereas mtDNAs of the unisexuals are closely related or indistinguishable from those of only one of the sexual progenitors. Thus, an emerging generalization is that most extant unisexual biotypes originated through asymmetrical hybridization events, occurring in one direction only (e.g., female

A × male *B* versus female *B* × male *A*). In many cases, mtDNA analyses have further pinpointed the geographic and genetic maternal source of the unisexuals. For example, nine unisexual biotypes in the *sexlineatus* group of *Cnemidophorus* lizards all appear to stem from females within one of the four nominate geographic subspecies (*arizonae*) of *C. inornatus;* and five triploid unisexual strains of *Poeciliopsis* in the *monacha-lucida* complex trace phylogenetically to an mtDNA haplotype observed in extant bisexual *P. monacha* from the Rio Fuerte in northwestern Mexico.

As shown by Spolsky and Uzzell, one exception to such straightforward hybrid origins involves the hybridogenetic frog *Rana esculenta,* in which individuals exhibit mtDNA genotypes normally characteristic of either *R. lessonae* or *R. ridibunda*. *Rana esculenta* is unique among the assayed "asexual" biotypes in consisting of high frequencies of both males and females. From behavioral considerations, the initial hybridizations producing *R. esculenta* have been postulated to involve male *R. lessonae* × female *R. ridibunda*. Once the hybridogen was formed, occasional matings of male *R. esculenta* with female *R. lessonae* secondarily may have introduced *R. lessonae*-type mtDNA into *R. esculenta*. Furthermore, females belonging to such *R. esculenta* lineages appear to have served as a natural bridge for interspecies transfer of *R. lessonae* mtDNA into particular *R. ridibunda* populations via matings with *R. ridibunda* males. Such crosses apparently produced "*ridibunda*" frogs with normal nuclear genomes (because the *R. lessonae* chromosomes are excluded during meiosis), but *R. lessonae*-type mtDNA.

Another complex scenario surrounds the hypothesized maternal ancestry of the triploid salamander *Ambystoma 2-laterale-jeffersonianum,* which by allozyme evidence contains nuclear genomes of the bisexual species *A. laterale* and *A. jeffersonianum,* but reportedly carries mtDNA from *A. texanum*. Kraus and Miyamoto favor an explanation in which an original *A. laterale-texanum* hybrid female produced an ovum with primarily *A. laterale* nuclear chromosomes, but the female-determining sex chromosome (W) and the mtDNA of *A. texan-*

um. When fertilized by a male *A. laterale,* female progeny with two *A. laterale* nuclear genomes and the mtDNA of *A. texanum* would result. Subsequent hybridization with male *A. jeffersonianum* could then produce the observed *A.* 2-*laterale-jeffersonianum* biotypes carrying *A. texanum* mtDNA. Although this scenario remains speculative, its mere feasibility suggests that distinct reticulate histories could characterize different genomic elements in some hybridogens.

All other unisexuals examined to date appear to lack the sorts of peculiarities associated with hybridogenetic reproduction in the *Ambystoma* and *Rana* amphibians, and interpretations of maternal ancestry are more straightforward.

Formation of Polyploids

About 64% of the known unisexual biotypes are polyploid. It has been suggested that normal meiotic processes are disrupted in interspecific hybrids, such that triploids (for example) might have arisen through a hybrid intermediate that produced unreduced diploid eggs subsequently fertilized by haploid sperm (henceforth the "primary hybrid origin" hypothesis). Alternatively, under the "spontaneous origin" hypothesis, parthenogenetic triploids might have arisen when unreduced oocytes from a diploid nonhybrid were fertilized by sperm from a second bisexual species.

mtDNA assays permit a test of these competing hypotheses. If a unisexual biotype arose spontaneously from sexual ancestors and hybridization was involved only secondarily, the paired homospecific genomes should derive from the maternal parent, and thus should be coupled with mtDNA from the same species (i.e., the AA genome of AAB should be coupled with mtDNA type *a*, and vice versa for ABB). Conversely, under a primary hybrid origin, the paired homospecific genomes could be coupled either with mtDNA type *a* or *b* (depending on details by which a nuclear genome was duplicated or added—see below).

Genetic analyses by Moritz and colleagues have provided strong support for the "primary hybrid origin" hypothesis for eight of ten parthenogenetic *Cnemidophorus* lizards. For example, the triploid parthenogen *C. flagellicaudus* possesses the mtDNA of *C. inornatus* but two homospecific nuclear genomes from *C. burti*. Similarly, as shown in our own lab, the triploid gynogenetic fish *Poeciliopsis monacha-2 lucida* possesses the mtDNA of *P. monacha* but two nuclear genomes from *P. lucida*. In these cases, the genetic data clearly refute the "spontaneous origin" scenario, and support the postulated relationship described above between hybridization, unisexuality, and polyploidy in vertebrates.

Assuming the correctness of the primary hybrid origin scenario, two further cytogenetic pathways to triploidy can be distinguished. Under the "genomic addition" scenario, interspecific F_1 hybrids produce unreduced allotriploid biotypes AAB or ABB. Under the "genomic duplication" scenario, suppression of an equational division in an F_1 hybrid could produce unreduced AA or BB ova, which upon backcrossing to species A or B would produce AAB or ABB offspring (autopolyploid AAA or BBB progeny could also result from this process, but no self-sustaining populations of autopolyploid unisexual vertebrates have been found). An important distinction between these pathways involves the predicted level of heterozygosity at loci in the homospecific nuclear genomes. Heterozygosity should be extremely low under the genome duplication pathway (the only variation being derived from postformational mutations), whereas normal heterozygosity is predicted under genomic addition. In triploid *Poeciliopsis* gynogens, it turns out that all assayed strains are heterozygous for homospecific nuclear markers at one or more allozyme loci, a result that effectively excluded the genome duplication hypothesis for these fishes.

Genetic Diversity within Unisexual and Bisexual Taxa

Estimates are available of mtDNA genotypic and nucleotide diversities within many unisexual taxa and the extant sexual descendants of their sexual maternal progenitors. Two major points are evident from an examination of these findings. First, mtDNA variability has been detected within most unisexual taxa. For example, several such "species" show genotypic diversity values greater than 0.50, indicating that random pairs of individuals are distinguishable with high probability, even under these limited genetic assays, which typically surveyed about 2% of the mtDNA genomic sequence. Second, mtDNA variation (nucleotide diversity) within the unisexuals is normally much lower than that within sexual relatives. Of 13 unisexual taxa studied (excluding *R. esculenta* because its mtDNA derives from two ancestral species), 12 showed mtDNA diversities less than or equal to those of their maternal sexual parents. Some comparisons were dramatic. For example, nucleotide diversities in the unisexuals *Cnemidophorus uniparens, Heteronotia binoei,* and *Ambystoma laterale-texanum* were more than an order of magnitude lower than such values in their sexual counterparts. Only in *P. monacha-lucida* did nucleotide diversity in a unisexual exceed (slightly) that of its sexual cognate.

Nevertheless, such prima facie comparisons are of limited value for the following reasons. First, genotypic and nucleotide diversity estimates can be strongly influenced by the sampling design and geographic distributions of genotypes, both of which varied greatly among studies. Second, taxonomic assignments can exert an overriding influence on interpretations of genetic diversity estimates. For example, the unisexual *Cnemidophorus tesselatus* exhibits nucleotide diversity essentially identical with that of its sexual cognate *C. marmoratus,* yet *C. marmoratus* is sometimes treated as a subspecies of the more widespread and variable *C. tigris.* If *C. tigris* as a whole were considered the bisexual ancestor of *C. tesselatus,* the estimate of mtDNA variability within the sexual form would be vastly greater than that within the unisexual. Taxonomic conventions applied to unisexual biotypes can similarly influence perceptions of genetic diversity. For example, nine unisexual "species" of *Cnemodophorus* lizards that had *C. inornatus* mothers and *C. burti* (or *C. costatus*) fathers are given taxonomic recognition on the basis of morphological or karyotypic distinctions, whereas all diploid *Poeciliopsis* unisexuals with *P. monacha* mothers and *P. lucida* fathers are conventionally considered a single fish taxon, *P. monacha-lucida,* despite differences among strains in morphology, ecology, and behavior.

Phylogenetic Relationships of Unisexual and Bisexual Taxa

In assessing the significance of mtDNA variability to questions of unisexual origins and ages, an important step is to distinguish between unisexual lineages that arose only once from those that had multiple hybrid origins. Toward that end, relationships in the matriarchal phylogenies of unisexual-bisexual complexes have been assessed. With respect to maternal phylogeny, three categories of relationship are possible between a unisexual biotype and its maternal sexual cognate: (1) *reciprocal monophyly,* in which all mtDNA lineages within the sexual species are more closely related to one another than to any lineages within the unisexual, and vice versa; (2) *paraphyly,* in which all mtDNA lineages within the unisexual are more closely related to one another than to any bisexual, but some lineages in the sexual species are more closely related to those in unisexuals than to one another (the converse of this direction of paraphyly is also conceivable); and (3) *polyphyly,* in which extant lineages of neither the unisexual nor the bisexual form a distinct clade.

These phylogenetic categories should be a function of the mode of origin of the unisexuals and of demography-based processes of maternal lineage sorting in the unisexual and sexual taxa. For example, a bisexual that gave rise to a unisexual via a single hybridization event would initially exhibit a paraphyletic mtDNA relationship to the unisexual (or in other words, the unisexual would constitute a monophyletic lineage within the broader matriarchal phylogeny of the sexual); through time, maternal lineage sorting within the sexual species might then convert the relationship to one of reciprocal monophyly. However, a unisexual that arose through multiple hybridization events involving unrelated females would initially appear polyphyletic in mtDNA ancestry. Maternal lineage turnover within the sexual and unisexual forms could subsequently lead to the appearance of paraphyly and then reciprocal monophyly in the mtDNA tree.

The phylogenetic status of mtDNA has been studied in at least 14 unisexual-bisexual complexes. Five such complexes exhibit mtDNA polyphyly, and these provide prima facie evidence for multiple, independent hybrid origins of unisexuals from crosses involving distantly related female ancestors. For example, from the mtDNA phylogeny of extant lineages my collaborators and I were able to infer at least five independent origins for diploid *P. monacha-lucida;* that is, genotypes in the unisexuals trace to at least five distinct nodes in the mtDNA phylogeny of the sexual ancestor *P. monacha.* Similar documentations of multiple hybrid origins occur in *C. tesselatus, Menidia clarkhubbsi, Phoxinus eos-neogaeus,* and, for the biological reasons discussed earlier, *R. esculenta.*

The remaining unisexual-bisexual complexes exhibit a paraphyletic status, in which the unisexuals form a mtDNA clade within the broader matriarchal phylogeny of the sexual ancestor. The evolutionary depths in the bisexual matriarchal phylogeny were often far greater than those within the unisexual derivatives, as indicated by the mtDNA phylogenies themselves, and by the maximum observed sequence divergences within the sexual versus unisexual taxa. As discussed, below, such instances of bisexual-unisexual paraphyly are consistent with but do not prove that single hybridization events were involved in the formation of these unisexuals.

The Number of Formational Hybrid Events

Each particular mtDNA clade within a polyphyletic or paraphyletic phylogeny discussed above appears to have arisen from hybridization event(s) involving one or more closely related bisexual females. Nonetheless, an important distinction should be drawn between unique mtDNA origin and individual hybridization event, and this can be seen most forcefully by noting that hybridization of even a single female with more than one male could produce multiple unisexual lineages differing in nuclear genotype but identical in mtDNA. In some cases, different nuclear genotypes, as identified by allozymes, morphology, or results of tissue grafts, have indeed been observed within

a mtDNA clone or clade in a unisexual. For example, about 12 different diploid allozyme genotypes have been found in populations of *C.
tesselatus* that from mtDNA evidence appear to have only two or three
distinct origins within the matriarchal phylogeny of the sexual progenitor, *C. marmoratus;* and each of several mtDNA clones within *P.
monacha-lucida* could be decomposed into distinct allozyme classes.
Conversely, several allozymically defined clones of *P. monacha-lucida*
could be subdivided into distinguishable mtDNA genotypes with the
restriction assays employed.

A serious complication in interpreting variability in nuclear or mitochondrial genomes concerns distinguishing postformational mutations from genetic differences frozen during separate hybrid origins.
Ancillary information may help in determining how many hybridization events were involved. For example, unique alleles that are rare
and localized in unisexuals probably arose through postformational
mutations, whereas distinct genotypic combinations that are common
in unisexuals and shared with extant sexual populations probably derive from independent hybridizations. Such reasoning suggests that at
least four of twelve allozyme genotypes observed in *C. tesselatus* originated through separate hybrid events. Similarly, most clonal diversity
in hybridogenetic *Poeciliopsis* fish also arose via multiple hybridizations, but postformational mutations resulting in silent enzymes,
deleterious recessives, and mtDNA variants marked some clones. In
other cases where the potential sexual progenitors of unisexual vertebrates are less well known and incompletely sampled, it is more difficult to discriminate between postformational mutations and frozen
variation. In general, although multiple hybridizations clearly were involved in the formation of several of the unisexual taxa, it is not yet
possible to specify the precise number of such events for any biotype.

Evolutionary Ages of Unisexual Biotypes

The relatively low mtDNA genetic variability within unisexual taxa
and the pattern of mtDNA paraphyly observed in most unisexual-bisexual complexes suggest severe constraints on the origin and/or sur-

vival of most asexual vertebrate lineages. These observations alone, however, do not establish the absolute evolutionary durations of surviving unisexuals.

One approach to estimating the age of an asexual lineage involves genetic comparison with the closest sexual relative. Such estimates are available for mtDNA sequence divergence between each of 24 putative unisexual clades and the closest haplotype observed in its respective bisexual progenitor. Of these 24 clades, 13 were indistinguishable in the assays from an extant genotype in the sexual species, indicating a very recent evolutionary separation; and five additional unisexual clades differed from nearest assayed bisexual at sequence divergence estimates that were less than 1%, suggesting times of origin within the past 500,000 years (assuming a "conventional" vertebrate clock calibration of 2% sequence divergence between a pair of lineages per million years). On the other hand, a few unisexual haplotypes showed greater sequence divergences, which translate into literal estimates of evolutionary origins as much as 2.75 million years ago. However, a serious reservation about such estimates is that closer relatives within the sexual progenitor may have gone extinct after unisexual separation, or otherwise remained unsampled in the collections. Indeed, because of the lower mtDNA diversity within unisexuals, most authors have concluded that unisexuals arose very recently, even when close mtDNA lineages were not observed among the sexual relatives sampled.

A second and more conservative approach to estimating evolutionary ages of unisexual taxa involves assessing the postformational spectrum of mtDNA variation within a unisexual clade that by *independent* criteria is thought to derive from a single hybridization event. This approach is less likely to produce artifactual overestimates of unisexual age, but on the other hand could seriously underestimate origination times because of postformational lineage extinctions within the unisexual clade. Few unisexual clades are well enough characterized to permit such mtDNA age assessments, but one case does stand out as exemplary of this method. Quattro and colleagues described postformational mtDNA diversity within a monophyletic lineage of *Poeciliopsis monacha-occidentalis* that, on the basis of independent zoogeographic evidence, protein-electrophoretic data, and tissue-grafting analysis, was of single-hybridization origin. From the observed nucleotide diversity within the clade, and by using a conventional mtDNA clock calibration for vertebrates, this unisexual lineage was estimated to be roughly 100,000–200,000 generations (ca. 60,000 years) old. Although this estimate is provisional due to uncertainties of the mtDNA evolutionary rate, an ancient origin is also indicated by postformational mutations observed at allozyme and histocompability loci, and by the 550-km geographic range of the lineage. Clearly, at least some unisexual vertebrate lineages can achieve considerable ecological success and evolutionary longevity.

Summary

Genetic surveys of mtDNA diversity have provided a novel class of information on the evolutionary histories of more than 25 "species" of unisexual vertebrates and their bisexual progenitors. A review of this literature reveals that (1) mtDNA inheritance in hybridogenetic fishes (and, by extension, other gynogenetic and parthenogenetic unisexuals) is indeed strictly maternal; (2) most unisexuals arose through nonreciprocal hybridization events between bisexual species in which the female parent has now been identified; (3) most polyploid unisexuals arose via fertilization of unreduced eggs in a diploid hybrid, rather than spontaneously via fertilization of eggs in a nonhybrid; (4) with few exceptions, overall mtDNA genetic diversity within an assayed unisexual taxon is considerably lower than that within its maternal bisexual cognate; (5) in terms of matriarchal phylogeny, a few bisexual-unisexual complexes exhibit a pattern of polyphyly (demonstrating independent hybridization origins of unisexuals from unrelated female ancestors), while most show paraphyly (in which cases the unisexual appears to have arisen only once or a few times from within a subset of the matriarchal genealogy of the sexual parent); (6) although many unisexuals are closely related to maternal lineages in the sexual relative, and thus appear evolutionarily young, in one well-characterized unisexual clade (involving *P. monacha-occidentalis*), numerous postformational mutations indicate an evolutionary age of at least 100,000 generations. Thus, contrary to conventional wisdom, some unisexual vertebrate clades can achieve considerable evolutionary longevity.

Unisexual biotypes provide exceptions to the norm of sexual reproduction in vertebrates. Mitochondrial DNA provides an exception to the norm of biparental Mendelian inheritance. These "aberrant" systems studied together have added a synergistic boost to our understanding of the evolutionary genetics of clonal systems.

7

Aging, Sexual Reproduction, and DNA Repair

Near the beginning of the 1990s, three influential books appeared on three seemingly different evolutionary topics: aging, sex, and DNA repair. In truth, all of these subjects are thoroughly intertwined (as the authors themselves realized). In the paper on which this chapter is based, Avise extended the connections among these topics by considering several additional factors that previously had been neglected, including the relevance of multicellularity versus unicellularity, and uniparental versus biparental inheritance patterns within the same organismal lineage. Thus, this chapter is partly a review of the three books listed above, and partly a synthesis with an entirely novel perspective. The original version appeared as the inaugural paper in a "Perspectives" section that ever since 1993 has been a regular feature leading off many issues of the journal *Evolution*.

Three seminal books on the evolutionary biology of aging and sexual reproduction appeared within a few years of one another: *The Evolution of Sex* (1988) edited by Richard Michod and Bruce Levin, *Aging, Sex, and DNA Repair* (1991) by Carol and Harris Bernstein, and *Evolutionary Biology of Aging* (1991) by Michael Rose. In this joint review of these three books, much of my own thinking on the topic is also incorporated. In particular, I pay close attention to the provocative suggestion by Bernstein and Bernstein that senescence and genetic recombination are related epiphenomena stemming from the universal challenge to life posed by DNA damages and the need for damage repair.

The phenomena of aging and of sexual reproduction are among the most counterintuitive and puzzling of widespread outcomes to have evolved under the influence of natural selection. Why should individuals of most species senesce and die when Darwinian selection seemingly would favor any genetic predisposition for greater longevity and continued reproductions? And why should individuals engage in sexual as opposed to asexual reproduction, when by so doing they not only expend time and energy in finding a mate, but also dilute (by 50%) their genetic contribution to each offspring? Evolutionary biologists have long pondered these issues, and the theoretical and empirical results were summarized eloquently in these three landmark books.

With his evolutionary perspective, Rose defines aging in his 1991 treatise as "a persistent decline in the age-specific fitness components of an organism (survival probability or reproductive output) due to internal physiological deterioration." The central thesis of Rose's book is that the mathematical framework of evolutionary genetics has solved the paradox of aging in age-structured populations by showing that the phenomenon is an inevitable outcome of the declining force of natural selection through successive age classes. Under the formal theory that Rose cogently summarizes, natural selection is simply indifferent to problems of somatic deterioration with advancing age, because as measured by effects on fitness (representation in successive generations) these problems are trivial compared with those that might appear earlier in life. Thus, aging and death exist not for any ineluctable physiological cause, but because of "a failure of natural selection to 'pay attention' to the problem." Particular genetic mechanisms of aging are not specified by this evolutionary theory, but two leading candidates for which explicit theoretical treatments are available are (1) antagonistic pleiotropy, in which alleles tend to evolve that have beneficial effects at early ages of life, but antagonistic deleterious effects later, and (2) age specificity of gene action, in which alleles with age-related deleterious somatic effects accumulate in evolution simply because they are nearly neutral in terms of fitness by virtue of weak selection in later age classes. Regardless of the means by which aging is played out from the basic evolutionary script, the take-home message is that "given age-structured populations and genetic variation in life histories, aging is a straightforward corollary of population genetics theory." This theory should apply to all organisms in which there is a clear distinction between somatic cells and germline cells.

Having established a conceptual primacy for the evolutionary theory of aging, Rose then chastises the field of gerontology for lack of this orienting foundation. For example, according to the evolutionary view, "the search for an ultimate physiological cause of aging is no more cogent than a search for a physiological cause of evolutionary adaptation would be . . . This implies that one of the basic goals of gerontology, that of finding the physiological cause(s) of aging, is misconceived." Rose provided extended reviews of the experimental evidence for several physiological theories for aging previously advanced (involving "wear and tear," rate-of-living considerations, hormonal influences, metabolic pathologies, and a host of others), and found all to be wanting as universal explanations. Although many of these factors no doubt play proximate roles in the aging process, none provides the ultimate explanation for aging that is embodied in the evolutionary view.

From experimental findings as well as comparative aspects of aging across life forms, Rose concluded that there are multiple causes for aging and that these can be arranged hierarchically with regard to explanatory power. The ultimate (evolutionary) cause is the attenuation of the force of natural selection with respect to the age of gene effects in species with soma. At the penultimate level are the population genetic explanations of antagonistic pleiotropy and mutation accumulation, and at the bottom tier are the highly idiosyncratic molecular, cellular, and physiological pathways by which the genetic underpinnings of aging happen to have been executed in a particular population or species.

Rose's book is a seminal contribution because it provides one of the clearest, most coherent and forceful documentations of why aging is not incompatible with natural selection after all. This new perspective should revolutionize the conceptual framework of gerontology, which as a discipline had remained one of the last bastions of biology that is relatively untouched by evolutionary thought. However, I do not quite share Rose's enthusiasm that this new theoretical orientation will revolutionize the day-to-day practice of gerontological research (any more than Darwin's classic, *On the Origin of Species,* changed the day-to-day practice of naming and describing species).

Thus, an important empirical task in gerontology will remain the identification of particular molecular or cellular events involved in the aging process, idiosyncratic as they may be. This effort is especially important in humans or other species in which ameliorative efforts might then be contemplated. Furthermore, if the arguments by the Bernsteins are correct (see below), then Rose's sounding of the death knell for global molecular mechanisms underlying aging may have been premature.

Sexual reproduction entails the generation of new combinations of genes by the mixing of genomes, or portions thereof. In most evolutionary definitions, *sex* is synonymous with *genetic recombination,* although some authors emphasize usual components of the process, such as *physical recombination* (the breakage and reunion of two different DNA molecules), and *outcrossing* (the mixing of DNA molecules from separate individuals). Why should various mechanisms for genetic mixis have evolved so nearly universally across life? Michod and Levin's edited book brings together authoritative and stimulating contributions on this topic from most of the major architects of recent theories on the evolutionary significance of sexual reproduction.

These diverse hypotheses can be divided into two categories that are nearly opposite in orientation, although not necessarily mutually exclusive. The first category of theories perceives a benefit per se for sex, either at the immediate level of individual fitness or at the evolutionary level of group persistence. Thus, genetic mixis itself is the object of selection. Theories of this type are united by the theme that genetic variability arising from mixis and molecular recombination must somehow be advantageous in an ecological or evolutionary theater, such that the benefits to individuals (or perhaps to extended groups) outweigh the rather obvious and substantial costs of sex to individuals. Three advantages classically proposed for sexual reproduction are as follows: (1) to facilitate the incorporation of beneficial mutations into an evolutionary lineage; (2) to facilitate the removal of deleterious mutations (i.e., to overcome Muller's ratchet, the ineluctable process by which the mutational load in strictly asexual lineages can remain only the same or increase through time); and (3) to allow adjustments to spatial or temporal changes in the physical and biotic environment. Several chapters (by Bell, Crow, Ghiselin, Maynard Smith, Seger and Hamilton, Williams, and others) formalize and elaborate these hypotheses, all of which can rationalize the prevalence of sexual modes of reproduction. However, some of the arguments are less than fully convincing, in particular, when it comes to the proposed short-term benefits of sex that are required under a strictly individual-selectionist framework.

The second category of theories proposes instead that sex is a coincidental evolutionary by-product of other primary consequences for mixis. For example, Hickey and Rose propose that sex is an outcome of subgenomic selection on parasitic DNA sequences that "imposed" biparental sexual reproduction on host genomes to favor their own spread. Another set of scenarios in this category (chapters by Bernstein et al., Holliday, Levin, and Shields) proposes that the evolution (and perhaps maintenance) of sexual reproduction involved selection pressure favoring mechanisms for the correction of genetic errors. This leads us finally to discussion of the *DNA-repair* theory of sex and aging, as further elaborated by Bernstein and Bernstein.

• • •

A fundamental tenet of the Bernsteins' theory is that damages to genetic material are a universal problem for life. These damages, defined as structural irregularities in DNA that cannot be replicated or inherited (unlike mutations), are of many types: single- and double-stranded breaks, modified bases, depurinations, cross-links, and so on. They arise inevitably from insults both endogenous and exogenous to the organism (e.g., oxidative damage from the molecular by-products of cellular respiration, and UV irradiation and DNA-damaging environmental chemicals, respectively). From empirical evidence, the cumulative numbers of such damages are astounding: for example, a typical mammalian cell experiences tens of thousands of DNA damages per day! These damages, if unrepaired, interfere with gene transcription and DNA replication and can cause progressive impairment of cell function and eventual cell death. The deterioration of somatic cellular function in turn leads to organismal senescence and death.

Damages to DNA can, however, be recognized and repaired by cells (although not necessarily at a rate that keeps up with their production). Enzymatic machineries for repair of DNA damages are evolutionarily widespread, and their molecular details have been worked out to varying degrees in several model organisms ranging from viruses and bacteria to mammals. DNA-repair processes almost invariably require the replacement of damaged genetic material through use of the intact information derived from a redundant copy. One source of redundancy is the complementary strand in double-helical DNA, which can serve as a template for repair when damage is confined to a single DNA strand. For example, all known forms of excision repair that occur regularly in somatic cells involve removal of the damaged section from one DNA strand and replacement by copying from the complementary undamaged strand.

A second source of redundancy for repair is the presence of another duplex DNA molecule with information homologous to that of the original copy. Such an undamaged template appears necessary for the *recombinational repair* of double-stranded DNA damage. The Bernsteins argue that the exchange of genetic information between multiple infective phages, as well as the process of transformation whereby some bacterial cells actively take up naked DNA from the surrounding medium, are examples of primary adaptations for DNA repair in these microbes. So too, they argue, is meiosis in higher organisms, which is viewed as an adaptation for promoting recombinational repair of the DNA passed on to gametes. In general, all mechanisms for molecular recombination are interpreted by the authors as evolutionary adaptations that originated *and* are actively maintained by natural selection explicitly for the functions they serve in recombinational repair of DNA damage. Furthermore, in diploid multicellular reproductive systems with recombination, the Bernsteins suggest that outcrossing is favored because it promotes the masking of deleterious *mutations*. Thus, "DNA damage selects for recombination, and mutation in the presence of recombination selects for outcrossing."

According to the DNA-repair theory, aging processes resulting from DNA damage should occur in all organisms, and not just those with a clear distinction between somatic tissues and germ-line cells. There appears to be a conflict of opinion (or perhaps merely a semantic distinction?) about whether senescence occurs in unicellular creatures such as bacteria, and in vegetatively reproducing multicellular creatures such as some plants and invertebrate animals. Rose concludes that "species that unequivocally lack such a separation of the soma, such as some sea anemones, some protozoa, and all known prokaryotes, appear to lack aging." However, the Bernsteins suggest that although populations of cells may survive indefinitely (e.g., in clonally reproducing trees and bacterial colonies), nonetheless "one would not expect to find old cells in a tree any more than one would find old cells in a growing culture of bacteria." To account for the persistence of such asexual populations of cells, the Bernsteins also introduce the concept of *cellular replacement,* in which lethally damaged cells are replaced by replication of undamaged ones. This strategy should work in any cell population in which "the incidence of unrepaired lethal damages is low enough at each generation to permit replacement of losses." Thus, the Bernsteins propose that there are two

possible pathways to immortality for a cell lineage: (1) recombinational repair of DNA damages (which applies to germ cells); and
(2) cellular replacement (which applies to predominantly clonal cells
as in many bacteria).

The joint pillars of the Bernsteins' theory are that aging is a direct
consequence of the accumulation of DNA damage, and that sex where
it occurs is a consequence of the need to transmit damage-free genetic information to progeny. The theory as presented does not imply
that the production of allelic variation through recombination and
outcrossing is unimportant for long-term evolution: "Infrequent beneficial allelic variants generated by recombination undoubtedly promote long-term evolutionary success, just as infrequent beneficial
mutations do." Nonetheless, "the tendency toward randomization of
genetic information that occurs with recombination and outcrossing,
under general conditions, has a negative effect on fitness in the short
run, just as mutations do."

The DNA-repair theory as expounded by the Bernsteins is extremely important for several reasons. First, it provides a conceptual
framework for linking the widespread phenomena of aging and sex,
two evolutionary subjects that more typically have been dealt with
separately (as in the Rose and Michod and Levin volumes). Second,
the theory appears both logically consistent internally, and eminently
plausible empirically—at least as much so as many of the traditional
theories on sex and aging. Indeed, much of the Bernsteins' book constitutes a detailed compilation of observations and experimental data
that appear either consistent with or positively supportive of the
DNA-repair view. Third, the DNA-repair theory envisions immediate
selective advantages that apply to individuals and their offspring and
not merely to longer-term group benefits.

Also, the DNA-repair theory represents a dramatic and refreshing
(to me) conceptual departure from the more traditional evolutionary
theories of sex, which sometimes seem to go to rather great lengths
in attempts to identify short-term benefits for the genetic variability
generated by recombination. Under the Bernsteins' view, genetic vari-

ability is an immediate curse rather than a blessing, with any long-term benefits derived from recombinational variation being fortuitous epiphenomena of cellular and molecular processes that evolved under selection pressures to repair DNA damages and mask deleterious mutations. In this regard, I am reminded of the opposing world views on genetic variation expressed in another evolutionary arena—the debate between the selectionists and the neutralists. When extensive genetic variation was first uncovered in protein-electrophoretic and other molecular assays, many evolutionists assumed that the variability must be actively maintained by natural selection, and they sought hard to identify the balancing selective forces involved. But from the neutralist perspective (which grew out of the "classical" school in which genomes were perceived as heavily burdened by mutational load), the overall magnitude of molecular variation was actually much lower than expected, given suspected mutation rates and effective population sizes. Thus, under the neutralist (and classicist) worldviews, if selection was involved appreciably in molding molecular genetic variability, it must act primarily in a diversity-reducing rather than diversity-enhancing fashion.

Where does the DNA-repair hypothesis fall within the hierarchical framework of causes for aging as advanced by Rose? If correct, the theory cannot be placed at the bottom of the hierarchy as just another idiosyncratic physiological mechanism for aging, because it is general, and an explicit selective force is involved. Indeed, the hypothesis is in some respects more universal than that of the declining force of natural selection with advancing age, because it applies to all forms of life, including those without a clear distinction between somatic and germ cells. However, for organisms with soma, the DNA-repair hypothesis does not appear incompatible with Rose's evolutionary view: the declining impact of natural selection with age would mean that any organismal benefits to accrue from DNA-repair processes in the later cohorts of an age-structured population would provide insufficient selective force to circumvent the evolutionary appearance of senescence and somatic death.

Having heartily applauded the Bernsteins' contribution, I must add, however, that I seriously doubt it tells the whole story on the significance of genetic variation. Once recombinational processes had evolved (for whatever reason, of which the need for DNA repair must now be considered a leading candidate), it seems probable that the genetic variability thereby generated would have been exploited for other functions as well. For example, the extensive molecular variability in the repertoire of the immune response in higher animals is in part recombinationally derived, and undoubtedly fosters enhanced disease resistance that often must be of immediate fitness benefit. Furthermore, the increased genetic variance stemming from recombination might well allow sexual reproducers to outpersist asexual reproducers in changing environments, despite the fact that such explanations tend to be group selectionist. Finally, as several authors emphasized in the Michod and Levin volume, rates and patterns of genetic recombination (and the linkage disequilibria that they entail) can vary remarkably across different regions of the genome, between the sexes, temporally within the life cycle (e.g., in taxa with an alternation of generations between sexual and asexual modes), across populations and species, and spatially across habitats. Many of these differences have been interpreted as adaptive adjustments to varying selection regimes. As Ghiselin stated in the Michod and Levin volume, "The eukaryotic genome turns out to be very highly organized, and the whole apparatus shows every indication that the amount, kind, and timing of recombination, and also the release of variability, are adaptive . . . the DNA repair hypothesis suggests that there should be little correlation between what goes on and when and where it happens. Such a correlation definitely does exist."

There are two major reasons why a relative neglect of mitochondrial (mt) genomes in these volumes was surprising (similar sentiments could also be expressed about chloroplast DNA). First, in organisms as diverse as fungi and humans, elsewhere there has been a tremendous resurgence of interest in the possible roles of mitochondrial DNA (mtDNA) damage in the aging process. In humans, for example, this interest has been prompted by empirical findings that specifiable defects in mtDNA accumulate with advancing age in somatic cells, and that these defects tend to compromise physiological functions particularly in tissues and organ systems with high energy demands (e.g., the central nervous system, optic nerve, heart and skeletal muscle fibers, kidney, and liver). These are also the organ systems commonly associated with degenerative disease and chronic illnesses of the elderly, thus suggesting a possible cause-and-effect relationship between mtDNA damage and the aging process.

Further empirical and conceptual reasons exist for postulating that mtDNA might play a disproportionate role in aging. Mitochondrial DNA molecules are housed in an intracellular environment where they would seem to be especially prone to damage from oxygen radicals generated by oxidative phosphorylation. Indeed, mammalian mtDNA receives about 16-fold more oxidative damage on a per-nucleotide basis than does nuclear DNA. Yet ironically, animal mitochondria are thought to possess only limited DNA-repair systems, and indeed this provides one conventional explanation as to why animal mtDNA evolves so rapidly at the nucleotide sequence level. Animal mtDNA is packed tightly with genes crucial to the energy metabolism of cells, and for this reason, too, it would seem highly desirable for organisms to have evolved refined mechanism for the repair of mtDNA damage. The paradox is heightened further because there are many copies of mtDNA within most cells. Thus it would seem that any repair capability should, in principle, be especially workable because of the presence of many templates against which DNA damages might be corrected. (The hypothesis that an immunity from selection pressures stems from mtDNA redundancy and a possible excess metabolic capacity seems gratuitous and is also probably untenable evolutionarily.) Perhaps eukaryotic organisms *have* evolved more highly refined mtDNA-repair mechanisms that, despite intensive searches, thus far have remained undiscovered. But if not, why not?

And, how can organisms have persisted evolutionarily without such enzymatic repair services for the crucial cytoplasmic genomes they depend on for energy supplies?

A second reason for surprise over the relative neglect of mtDNA in these books relates to mtDNA's asexual inheritance. The transmission of mtDNA in most higher eukaryotes is predominantly uniparental, with effective genetic recombination between maternally and paternally derived molecules essentially unknown. If meiosis and the recombinational aspects of gametogenesis provide evolutionary benefits, as surely they must (either via repair of DNA damages, and/or through generation of advantageous recombinational variation), then why doesn't mtDNA play by these rules? The entire answer cannot simply be that mitochondrial elements have been physically confined to the cytoplasm and hence unable to avail themselves of meiosis, because transfers and successful incorporations of some mitochondrial genes to nuclear chromosomes are known to have occurred over evolutionary time.

If meiosis is primarily a process for correcting DNA damages (as proposed by the Bernsteins), then mtDNA damages must be overcome by some process other than meiotic recombinational repair. One distinct possibility is that mtDNA molecules might occasionally undergo nonmeiotic recombinations (or gene conversion events) within the germ line, perhaps in such a way that damage-free mtDNA templates correct faulty ones. The relatively few experimental attempts to uncover physical recombination in animal mtDNA through use of genetic markers have been hampered by the usual predominance of only one or a few detectable mtDNA clones within most individuals. More intensive searches for mtDNA recombination are needed. Another possibility (elaborated beyond) is that processes of mtDNA replication and sorting during gametogenesis provide an alternative, nonrecombinational pathway for circumventing the accumulation of genetic damages.

...

Another issue that was underemphasized in these volumes concerns the evolutionary ramifications of varying degrees of cellular autonomy. The somatic cells of an individual usually are interdependent, both structurally and functionally, whereas gametes are relatively autonomous (except perhaps in rather "trivial" respects such as the collaborative efforts required of sperm in penetrating the eggs of some species). In other words, gametes tend to be cellular free agents, whereas somatic cells (in particular, in tightly organized creatures with determinate growth, such as many higher animals) are trapped in a web of interdependencies. James Crow in the Michod and Levin volume raised an important question: "Is passing through a single-cell stage itself important? . . . Starting with a single cell, sexual or asexual, permits each generation to begin with a tabula rasa largely unencumbered by the somatic mutations from previous generations." Crow went on to lament that "I have never heard the importance of going through a single-cell stage expressed before, and would welcome comments . . . as to its possible merits."

It seems to me that many of the fundamental distinctions commonly made in discussions of aging and sex—senescence versus immortality, sexual versus asexual reproduction, somatic versus germline tissue, unicellularity versus multicellularity, and individuals versus groups—are inextricably related, and might profitably be viewed through a common denominator revolving on the concept of cellular autonomy, as described next.

I propose some possible extensions to Bernsteins' theory of DNA repair, and by so doing suggest how concepts of cellular and molecular autonomy might usefully be added to future discussions on aging and sex. As mentioned above, two potential pathways to immortality seem available to life. The first is predominantly or exclusively asexual and is exemplified most clearly by unicellular organisms such as bacteria. Here, cell proliferation apparently can outstrip the rate of accumulation of DNA damages and deleterious mutations, with the net result that Muller's ratchet is circumvented and an indefinite continuation of the population occurs via cellular replacement. The second pathway is sexual and is exemplified most clearly by germ-cell lineages in multicellular organisms such as vertebrates. Here, repair of nuclear DNA damages by genetic recombination supposedly operates in conjunction with cell proliferation and intercellular selection to counter the accumulations of nuclear DNA damages and deleterious mutations that would otherwise be expected.

In both routes to immortality, many cells (bacteria or gametes) may die genetic deaths (e.g., from the inevitable imperfections of any DNA-repair mechanism), but these deaths do not compromise the continuance of cell lineages that happen to have escaped or repaired DNA damage. Thus, the efficacy of both pathways to immortality would seem to depend critically on the autonomy of the proliferating cells. To emphasize why this so, consider the prospect of somatic immortality for a multicellular organism such as a vertebrate. Even if some somatic cells and tissues could keep pace with DNA damage via the nonsexual strategy of cellular replacement (as may essentially be true for epithelial cells of the digestive tract of mammals, or for hemopoietic stem cells), these replacements are to no avail in conferring immortality because the final fate of these cell lineages remains inextricably tied to the remainder of the individual's soma (which as a whole inevitably senesces, as predicted by Rose's evolutionary theory). However, autonomous gametes and the genomes they contain *can* escape the sinking somatic ship.

This line of reasoning also illustrates the difficulty (semantically and otherwise) of disentangling the issue of immortality from that of the distinction between somatic and germ-line cells. Without the presence of somatic tissue, the evolutionary theory of Rose predicts no age structure in a population, and hence no aging; but without aging, there is no compelling evolutionary stimulus for the escape of autonomous cells from a soma that inevitably deteriorates (either from DNA damage or other causes). These ruminations also point out why the distinction between an individual and a population can become rather vague in discussion of aging and immortality in unicellular taxa. A bacterial colony may survive indefinitely, but without a distinction between somatic and germ cells, what is the organismal entity to which this immortality refers? In truth, what persists are certain cell lineages, but in this sense the "individuals" or "populations" are no more well defined than are the potentially immortal germ-cell lineages in higher taxa. Furthermore, many bacterial cells inevitably die genetic deaths, but without somatic benchmarks to assess chronological age, it is debatable whether this should properly be referred to as an "aging" phenomenon.

In many plants and invertebrate animals with various asexual modes of reproduction, the usual distinctions between individuals and populations, between somatic lines and germ lines, and between aging and immortality, all become ever more ambiguous. For example, vegetative cell lines of some plants can be maintained indefinitely (perhaps by the strategy of cellular replacement), whereas others appear to senesce (perhaps because cellular replacement cannot keep pace with DNA damage). The former might well be considered potentially immortal, but according to Rose they do not violate the evolutionary theory of aging because specification of germ-line tissue in these cases is problematic. Whether this is a definitional slight of hand or a bona fide consideration is unclear to me, but in any event a more critical factor may be the degree of cellular autonomy displayed. Diploid cells or collections thereof that have a capacity to survive and reproduce mostly independently of other cells exhibit considerable cellular autonomy (by definition). Thus, to a vegetatively spreading plant or coral, death of a portion of the "soma" may have relatively little influence on the survival and reproduction of the remaining cells of the genet (i.e., of a given clonal genotype, regardless of how it is physically partitioned). This contrasts with the situation in vertebrates, in which the death of a critical tissue dooms all somatic cells within each well-demarcated individual. Thus, any cell lineages characterized by increased levels of functional and replicative autonomy carry the potential for indefinite evolutionary persistence. Whether this potential could be realized then depends on additional factors, including whether the available processes of cellular repair and replacement are adequate to control DNA damages and to circumvent Muller's ratchet.

One important consideration on whether such cellular processes are workable indefinitely concerns genomic size. Formal models indicate that Muller's ratchet may well set an upper limit on the size of the genome in asexual organisms, especially when their populations are small. The small size of mtDNA molecules in higher animals (16 kilobases) may be a reflection of Muller's ratchet, and, furthermore, as noted by Graham Bell in the volume edited by Michod and Levin, the somewhat larger mtDNA molecules of yeast and plants "would have to recombine in order to maintain the integrity of their genomes, as seems to be the case." From this perspective, nuclear genomes are vastly too large for long-term effectiveness of a cellular proliferation strategy acting alone to compensate for accumulation of DNA damages and deleterious mutations, hence the additional requirements for sexual reproduction and recombination. Crow (in the Michod and Levin book) regarded this as an important factor accounting for why species with obligate parthenogenesis or other forms of asexual reproduction "are the twigs on the phylogenetic tree, not the main stems and branches."

I would like to propose that elements of both the recombinational repair and replacement strategies are employed simultaneously within the germ-cell lineages of higher organisms. Under this view, recombinational repair helps purge the nuclear genome of DNA damages, and a molecular-level analogue of cellular replacement ("molecular replacement") facilitates the purging of both DNA damages and deleterious mutations in nonrecombining cytoplasmic genomes. The immediate effect of these collaborative processes is to increase the probability that at least some gametes are produced that are free from genetic defects that had accumulated during the lifetime of the parent. In turn, the zygotes and early embryos produced by such "cleansed" gametes have a higher initial likelihood of being unburdened from the load of parental DNA defects.

The molecular replacement process is proposed to operate through the replicative segregation of mtDNA molecules in the lineages of germ cells (in particular, oocytes). Unlike nuclear genes in diploid organisms, each of which exists as a single allelic copy per gamete, thousands of mtDNA molecules populate most cells, and several hundred thousand copies may cohabit a mature oocyte. As cells undergo mitotic or meiotic cytokinesis, particular mtDNA mutations may fluctuate in frequency because of intracellular selection (differential replication) and genetic drift. Notably, the many mtDNAs in mature oocytes probably stem from a vastly smaller pool of mtDNA molecules that survive the process of replicative segregation in earlier cytokinetic divisions of the germ-cell lineage. Evidence for this conclusion comes from the empirical generality that most of the heterogeneity in mtDNA genotypes is distributed among rather than within individuals (implying relative mtDNA population bottlenecks in germ lines), and from observed rates of mtDNA clonal sorting in the gametes and progeny of heteroplasmic females. The net effect is that the mtDNA molecules that survive and replicate to populate a mature oocyte presumably have been rather scrupulously screened by natural selection for replicative capacity and functional competency in the germ-cell lineages they inhabit.

To the extent that these two damage-repair processes (recombinational repair of nuclear DNA and molecular replacement of cytoplasmic DNA) fail during gametogenesis, the metabolic functions of some germ cells will be compromised, and there will be gametic deaths. These gametic screening processes appear to have considerable scope and impact, for at least two reasons. First, germ-line cells are highly active metabolically, such that any functional defects likely would be exposed to cellular-level selection. Second, gametes are produced in prodigious quantities by most species (e.g., males produce billions of sperm, and the number of oocytes present in a human female at birth

is approximately 2,000,000). Furthermore, subsequent rounds of selective screening no doubt occur at the zygotic stage and during embryonic development, as genomes from the surviving functional gametes are called on to interact properly in diploid condition. Failures
at this level would be registered as embryonic abortions, which also
are known to occur with high frequency (e.g., the loss of all human
conceptions has been estimated at nearly 80%). In general, the Bernsteins interpret such observations to indicate that DNA damage is so
pervasive that "recombinational repair during meiosis, as well as
other repair and protective processes, may be just barely able to cope
with DNA damage."

The Bernsteins' DNA-repair theory by itself probably cannot account
for all of the variety and nuances of sexual reproduction and aging
processes. Nonetheless, it represents an exciting and important piece
of a jigsaw puzzle whose other elements are summarized so eloquently in the Rose and Michod & Levin books. Furthermore, in this
puzzle's emerging picture, aging and sex can be seen more clearly as
interrelated phenomena, both evolutionarily and mechanistically.
Undeniably, certain cell lineages in all extant life-forms have solved
the problem of innate mortality (at least over the four billion years of
life on Earth), and the strategies of genetic recombination, cellular replacement, and molecular replacement by which this has been accomplished are coming into sharper focus.

8

The Real Message from Biosphere 2

The Biosphere 2 facility in central Arizona began in 1991 as a privately financed pet project of Texas billionaire Ed Bass, but it quickly got a reputation in the media as something of a folly with regard to the science it produced. Accordingly, for a brief time in the 1990s an ad hoc scientific advisory panel was formed, of which Avise was a member. Unfortunately, the panel was convened only in the eleventh hour, and nothing much came of its last-ditch efforts to scientifically resuscitate the much-maligned facility. Nonetheless, as described in this chapter, the broader Biosphere experience may carry some important societal lessons about human–environment interactions.

On a September morning in 1993, eight gaunt but triumphant Biospherians emerged through the airlock doors of Biosphere 2 after two years under public scrutiny and sealed glass. Their re-entry into Biosphere 1 (Earth) marked completion of the first in a century-long series of planned missions, the stated objectives of which are to explore scientific frontiers in ecotechnology (for better husbandry of the planet's resources and as a model for colonizing space) and, in general, to inspire the human spirit. The latter goal may have been achieved. Aficionados see the endeavor as audacious and visionary—"the most exciting venture undertaken in the U.S. since President Kennedy launched us towards the moon," according to one excited commentator. And, unlike NASA's lunar mission, this $150 million program was launched entirely from private venture capital!

For those who don't know, or may have long forgotten, Biosphere 2 is a futuristic glass and steel "greenhouse" nestled in Arizona's Sonoran desert, about 30 miles north of Tucson. Engineered to be a self-sustaining mesocosm, almost completely sealed off from atmospheric or other material exchange with the outside world, the graceful three-acre enclosure housed nearly 4,000 introduced species of plants and animals in a Garden-of-Eden-like setting of tropical rainforest, marsh, desert, savannah, streams, agricultural area, and even a miniature ocean complete with coral reef. Biosphere 2 receives energy as sunlight and as electricity (from an adjacent natural-gas power plant) that drives a vast "technosphere" of pumps, sensors, scrubbers, air-cooling systems, and other electronic and engineering wizardry designed to keep the environmental systems within boundaries suitable for life.

I visited Biosphere 2 as an independent researcher, and I have to admit it set my mind aspin with ambivalent impressions. There was the commercial side—on adjacent grounds you could purchase biomeburgers, habitat hotdogs, and planetary pizzas, or browse gift shops and bookstores. There was also a mystical side, exemplified by the many evocative sculptures with names of Indian Gods fashioned of stainless steel salvaged from the Los Alamos atomic bomb project. There was the educational side, where thought-provoking films and tours explained ecosystem functions and their relevance to the design of space modules.

There were also the many ecotechnological paradoxes of Biosphere 2 itself, where earthly smells of compost and forest contrast with the electronic sterility of the computer control room, and where the Biospherians' simple agrarian lifestyle seemed in opposition to their sophisticated telecommunications with the international press. And then there was the scientific side, a focus of much controversy and media attention. Whether sound basic research happened or could ever find a good home in Biosphere 2 is debatable.

But the overriding scientific lessons from Biosphere 2 already may be available. To many of us, healthy ecosystems and biodiversity on Biosphere 1 (the Earth) have inestimable aesthetic value. But the economic legacies of three billion years of evolution are sometimes easy to overlook. Some broad-thinking economists have sought to attach dollar values to natural ecosystems by virtue of the fundamental life-support services rendered (e.g., atmospheric regulation by rainforests and oceans, water purification by marshes, groundwater storage by aquifers, soil generation and maintenance by decomposers), but such attempts are almost hopelessly complicated by the vast range of spatial and temporal scales over which the monetary valuations might be tabulated. However, thanks to the controlled experiment of Biosphere 2, we now have a more explicit ledger.

The cost of the man-made technosphere that (marginally) regulated life-support systems for eight Biospherians over two years was about $150,000,000 total, or $9,000,000 per person per year!

During their two years of voluntary incarceration, the Biospherians became acutely aware of their intimate connections with, and complete dependence on, the fragile ecosystems within Biosphere 2. As one Biospherian later wrote, "It seemed as though we had touched every aspect of our world; we interacted with molecules and with trees, we knew our environment's boundaries and its subtleties." The Biospherians would never have tolerated in their small household the kinds of practices that are so widespread in our broader world—massive deforestation, water and atmospheric pollution, the dumping of toxic chemicals, or over-exploitation of renewable and nonrenewable resources. Nor would human population growth within Biosphere 2 have been tolerable—both oxygen and food supplies already were stretched to the limits, to the point where supplemental oxygen had to be injected at the end of year one, and the scanty food stores had to be placed under lock-and-key to prevent recurring incidences of theft by the hungry Biospherians. Clearly, the facility was close to if not well beyond human carrying capacity.

Exactly how many people the Earth can hold remains uncertain, but many signs indicate that we are rapidly approaching achievable limits. Indeed, if carrying capacity is defined (as it often is) as the maximum population that can be supported without degrading the environment, then the Earth's carrying capacity already has been exceeded. Ozone depletion and atmospheric pollution are global concerns, as are losses of groundwater supplies and usable surface waters, soils, fossil fuels, and species. Massive hunger, starvation, and conflicts over limited resources are recurring themes in many regions of the world. Current population densities over vast areas of the planet are not grossly different from those in the crowded Biosphere 2. Even if saying it is repetitious, it still strikes me as astonishing that our species currently shows a net increase of more than 10,000 people every hour, a quarter million people each day. How much farther the Earth's life-support systems can be pushed remains to be seen, but all of us are unwitting guinea pigs in this reckless and utterly pointless experiment with global carrying capacity. Unlike the inhabitants of Biopshere 2, we have no outside source of rescue or escape. We can only save ourselves, through humane efforts at population control.

Herein lays the real message Biosphere 2. It may be fun and even inspirational to dream of colonizing other planets, but the harsh reality is that we have only one home and it is getting crowded. Like the astronaut's views from space, Biosphere 2 should give us a novel perspective and renewed appreciation of Biosphere 1. It took vast time for evolution to design the Earth's biotas and ecosystems, and, like many people, I can't help but wonder if our disregard for this long evolutionary journey will be our undoing. Whether based on ethical or purely utilitarian considerations, human societies must learn to value our Earth properly, and quickly.

9

Conservation Genetics and Sea Turtles

Marine turtles are wonderful subjects for genetic analysis because molecular markers can uncover many aspects of their behaviors and evolutionary histories that are otherwise difficult or impossible to address due to the animals' oceanic habits and long life spans. All living species of sea turtle are endangered or threatened, so these charismatic reptiles are also of special conservation concern. In 1997, Avise was asked to deliver the Wilhemine E. Key Invitational Lecture of the American Genetic Association to kick off a special symposium devoted to conservation genetics in the sea. For several years, students and collaborators in his laboratory (Brian Bowen, Steve Karl, and others) had been studying sea turtles using various classes of molecular marker. So, the Key Lecture gave the author an opportunity to review genetic findings on sea turtles, and to place these discoveries in the broader context of conservation efforts in the world's oceans. This chapter presents relevant excerpts from that review.

If not for the fact that about 70% of the earth's surface is covered by oceans, the long-term prospects might be even dimmer for the biosphere's eventual recovery from global environmental crises precipitated by human overexploitation. The oceans have resisted permanent human settlement, and their vast size and composition provide some buffer against global environmental insults by man. Yet, even in the relatively untouched marine realm, human impacts on biodiversity have been profound. Populations of many of the world's largest and most spectacular marine mammals, reptiles, birds, fishes, and invertebrates have been depleted severely or forced to extinction by human harvesting, by human activities that pollute or otherwise degrade saltwater environments, or from the effects of human-mediated introductions of alien species.

Conservation issues for marine organisms have attracted the attention of many geneticists. The explosion of interest in conservation genetics was made possible by the deployment in the past three decades of usable laboratory techniques for the direct assay of DNA and proteins. Prominent among these have been mitochondrial DNA (mtDNA) assays that permit a characterization of matrilineages within and among species, various nuclear assays (e.g., of allozymes or microsatellite DNAs) that yield genotypic descriptions for particular Mendelian loci, direct nucleotide-sequencing methods that in principle can be applied to any nuclear or cytoplasmic genes, and polymerase chain reaction (PCR) procedures that permit recovery of DNA from even tiny amounts of tissue. These molecular procedures have opened the entire biological world for genetic scrutiny. Molecular techniques permit genetic analyses that were unimaginable earlier in the century, when the primary access to information on particular genetic traits came either from captive pedigrees or (indirectly and insecurely) from morphological and other organismal appraisals.

Sea turtles represent many of the traits present in species requiring conservation efforts in the marine realm. During its lifetime a sea turtle experiences dramatic growth, covers vast geographic areas, and is extremely vulnerable to human impacts. A long-standing question in marine turtle research has been whether females, after a sexual maturation process measured in tens of years and oceanic movements often measured in thousands of kilometers, return to nest at or near their natal beaches. Decades of field observations and physical tagging experiments have failed to answer this question conclusively. However, several pioneering molecular surveys of mtDNA have recently shown that conspecific rookeries of green turtles (*Chelonia mydas*), loggerheads (*Caretta caretta*), and hawksbills (*Eretmochelys imbricate*) within an ocean basin commonly display large or nearly fixed differences in matriline frequencies, a result that strongly supports natal homing scenarios for adult females. Because females ultimately govern the reproductive output of a rookery, this natal-philopatric behavior signifies a considerable demographic autonomy of each turtle rookery with regard to its reproduction. Thus, natural recruitment from foreign rookeries is unlikely to compensate for mortality in heavily exploited rookeries or to reestablish (over ecological timescales) rookeries that have been extirpated by human activities or other causes.

Molecular genetic markers also have been used to decipher movement and association patterns of marine turtles at other stages of the life cycle. Marine turtles spend most of their lives on oceanic journeys or on feeding grounds that may be far removed (hundreds or even thousands of kilometers) from rookery sites. Several studies have employed rookery-characteristic mtDNA markers to assign individuals captured on feeding grounds or during migration to rookeries of origin. An emerging generality is that particular assemblages of nonnesting marine turtles often derive from multiple rookery sites. Thus, with regard to mortality sources at nonnesting phases of the life cycle, different rookeries can be jointly impacted demographically. This too can have conservation ramifications.

For example, the shells of hawksbill turtles are highly prized for "tortoiseshell" jewelry and ornamental products. Although a moratorium exists on international trade in hawksbill shells, in 1992 Cuba announced its intent to resume hawksbill harvests within its territorial waters. Genetic analyses of mtDNA from a nearby feeding population (Mona Island, Puerto Rico), and from several nesting colonies throughout the Caribbean, demonstrated that hawksbill turtles within a feeding assemblage can derive from multiple rookeries across a broad area. By logical extension, Cuban harvests of hawksbill turtles within its sovereign waters might be expected to have demographic impacts on multiple rookeries beyond its own.

Articles 66.1 and 67.1 of the *United Nations Convention on the Law of the Sea* prescribe that those countries that provide developmental habitats for particular species hold primary interest and responsibility for conserving those stocks. Marine turtles spend most of their lives far removed from natal sites, and thus are routinely exposed to harvest by foreign States. Mentioned above was the genetically estimated rookery composition of hawksbill turtles on a Puerto Rico feeding ground. Other examples in which mtDNA markers have identified the rookery sources of marine turtles killed at sea include, for example, the assignment of more than 50% of juvenile loggerheads in a Mediterranean longline fishery to nesting beaches in the Americas, and assignments of nearly all loggerhead turtles captured in North Pacific driftnet and longline fisheries to rookeries in Japan. These genetic assignments were made possible by large frequency differences in mtDNA haplotype between rookeries, and by the availability of statistical procedures (originally developed for mixed-stock fisheries) to quantify relative genetic contributions from multiple source populations. Apart from the extraordinary migrational feats for marine turtles documented by these genetic studies, the findings also raise jurisdictional questions related to the United Nations Articles. For example, do the nations whose endangered turtle populations are impacted by high-seas fisheries have the legal right and/or obligation to seek and enforce conservation agreements with harvesting nations?

The spatial scales of organismal dispersal and population structure in the sea often are vastly greater than those typifying most terrestrial animals. Many marine species are actively vagile as adults (e.g., pelagic turtles, fishes, and cetaceans), passively mobile over huge distances (e.g., zooplankton), or highly dispersive as gametes or larvae (including corals and mollusks that may be sessile or demersal as adults). Exceptionally high dispersal potentials sometimes translate into minimal or modest phylogeographic divergences over vast areas. For example, several billfish and tuna species show ocean-wide or even circumglobal levels of mtDNA differentiation that are no greater than those reported among populations of terrestrial vertebrates or freshwater fishes within small continental regions such as the southeastern United States. Thus, a special logistic challenge in population genetic studies of marine taxa is to conduct molecular surveys at spatial scales (sometimes global) commensurate with their possible population genetic patterns, given the organisms' dispersal potentials and any plausible historical connections between the water masses they inhabit.

On the other hand, a growing appreciation from studies of marine organisms is that high dispersal potential often does not translate into high levels of realized gene flow (even as registered in presumably neutral molecular markers). Many examples have come to light in which a population genetic subdivision in a marine species is pronounced despite high intrinsic organismal vagility. Such population structure may result in part from behavioral philopatry during the life cycle (as of female sea turtles and anadromous salmon); social organization into kinship groups (as in some cetaceans); or habitat restrictions and historical or contemporary physical partitions of suitable marine environments (as in sardines and anchovies). The magnitudes as well as the ecological and evolutionary processes responsible for a realized population genetic structure nearly always bear direct relevance to any conservation or management plans for the particular species involved.

Varying temporal depths of population genetic structure are of course possible within a species. To acknowledge this fact in a conservation context, biologists often distinguish "evolutionarily significant units" (ESUs; relatively deep historical population subdivisions) from "management units" (MUs; shallower but nonetheless differentiable population segments connected by little or no contemporary gene flow). My colleagues and I have suggested formal empirical guidelines for genetically identifying intraspecific population segments that should merit recognition as ESUs. These center on four distinct aspects of genealogical "concordance": (1) in the particular population units demarcated by multiple sequence characters (e.g., nucleotide substitutions) within a nonrecombining segment of DNA; (2) in significant genealogical partitions across multiple independent (unlinked and nonepistatic) loci; (3) in the geographic positions of intraspecific gene-tree partitions across multiple codistributed species; and (4) between gene-tree partitions and historical geographic boundaries as inferred from traditional (i.e., nonmolecular) biogeographic evidence.

Studies on marine turtles exemplify nicely the distinction between ESUs and MUs, and also illustrate how both can be relevant to population stewartry and conservation. Consider, for example, global phylogeographic patterns in mtDNA displayed by green turtles and loggerheads. As already mentioned, conspecific nesting rookeries of both species often show highly significant differences in matriline frequencies within ocean basins, and therefore qualify as MUs. However, these genetic differences typically are shallow with respect to the magnitudes of sequence divergence that distinguish the rookery-specific mtDNA haplotypes. By contrast, rookeries from separate ocean basins, notably the Atlantic-Mediterranean versus the Indian-Pacific, usually show much larger mtDNA sequence differences. Furthermore, in both of these turtle species, the inferred times of separation based on a testudine-specific molecular clock are in general agreement with plausible historical population separations prompted by the rise of the Isthmus of Panama some three million years ago. Thus, as gauged by concordance criteria (1), (3), and (4), with only minor exceptions, the Atlantic-Mediterranean rookeries within each species empirically constitute one ESU and the Indian-Pacific rookeries constitute another. Such ESUs are important because they represent relatively deep historical genetic subdivisions, within a species, that should warrant special conservation recognition. Individual rookeries are important also, as MUs, because their matrilineal differences imply that they are demographically independent (over ecological timescales) with regard to reproduction.

Discussions of phylogeographic population structure often grade into deliberations about systematics, taxonomy, and conservation prioritization. The marine turtles again provide illustrations. In the eastern Pacific, a dark-colored form of the green turtle sometimes has been afforded taxonomic recognition as a distinct species, the black turtle (*Chelonia agassizi*). However, in terms of placement within the global mtDNA phylogeny for *C. mydas,* black turtles proved to be essentially indistinguishable from other members of the Indian-Pacific green turtle clade. So, they might better be termed merely a color morph, or perhaps a subspecies.

Similar molecular studies of another complex of marine turtles have provided a contrasting outcome. The Kemp's ridley turtle (*Lepidochelys kempi*) was highly suspect taxonomically because of its near morphological identity to the olive ridley (*L. olivacea*), and because of an unusual range distribution that at face value made little biogeographic sense. The Kemp's ridley originally was described from a single nesting location near Tamaulipas, Mexico in the western Gulf of Mexico, whereas rookeries of the olive ridley occur nearly worldwide in suitable waters. Nonetheless, a molecular survey of mtDNA revealed that populations of *L. olivacea* from the Atlantic and Pacific Oceans are considerably less differentiated from one another than either is from *L. kempi,* and that the Kemp's ridley is slightly more distinct on average from these olive ridleys than are *any* conspecific populations of green turtles or loggerheads to one another. In this case the genetic results bolstered the biological rationale for taxonomic recognition of the Kemp's ridley, and, thus, for the focused international conservation efforts that had been directed toward it.

Closing Thoughts

Traditionally, the field of conservation genetics was preoccupied with levels of genetic variation and possible inbreeding depression in small, often captive animal populations. However, with the advent of molecular approaches and their suitability as sources of polymorphic genetic markers for natural populations, a host of additional issues of conservation relevance can be addressed, ranging from studies of organismal behaviors, natural histories, and population demographies, to assessments of spatial and temporal aspects of population structure, to elucidations of systematics and phylogeny at any scale.

Eliot Norse recently lamented that "Marine conservation biology lags terrestrial conservation biology by about 20 years. It lacks central paradigms, graduate training programs, and substantial dedicated funding." On the other hand, it is also true that the marine realm has provided some of the most exciting, imaginative, and innovative of available case studies in all of conservation genetics.

10

The History and Purview of Phylogeography

The journal *Molecular Ecology* was inaugurated in 1992 and soon became widely recognized as a leading scientific outlet for studies involving molecules and natural history. In 1998, a special issue of the journal was devoted to phylogeography, which by that time had become a popular branch of population genetics and phylogenetics. Avise introduced the 1998 issue of *Molecular Ecology* with this personal reflection on the birth and growth of phylogeography as a recognizable discipline. (A far more complete description of the field can be found in the author's book *Phylogeography: The History and Formation of Species*, 2000; an extended personal account of Avise's life and career in molecular ecology, phylogeography, and evolution is available in his autobiography, *Captivating Life: A Naturalist in the Age of Genetics*, 2001)

Phylogeography as a formal discipline was christened in 1987. However, the field's gestation and birth occurred in the mid-1970s with the introduction of mitochondrial DNA (mtDNA) analyses to population genetics, and to the profound shift toward genealogical thought at the intraspecific level (now formalized as coalescent theory) that these methods prompted. Phylogeography is a field of study concerned with the principles and processes governing the geographical distributions of genealogical lineages, especially those that occur within a species (i.e., intraspecifically). Use of the word phylogeography in the evolutionary genetics literature has grown exponentially since 1987, and literally thousands of papers have employed "phylogeography" in the title or as an index word.

As a subdiscipline of biogeography, phylogeography emphasizes historical aspects of the contemporary spatial distributions of gene lineages. The analysis and interpretation of lineage distributions usually requires input from molecular genetics, population genetics, phylogenetics, demography, ethology, and historical geography. Thus, phylogeography is an integrative discipline.

In purest form, empirical phylogeographic analyses deal with the spatial distributions within and among populations of alleles whose phylogenetic relationships can be deduced. Because mtDNA evolves rapidly in populations of higher animals and usually is transmitted maternally without intermolecular recombination, it has been the workhorse of most (>80%) of phylogeographic studies. However, empirical or theoretical treatments that address phylogenetic aspects of the spatial distributions of any genetic traits (morphological, behavioral, or any other) also can qualify as phylogeographic under a broader definition of the term. Furthermore, a matrilineal phylogeny (or any other allelic transmission pathway) constitutes only a minuscule fraction of the composite genealogical information within a sexual pedigree. A phylogeny for spatially structured populations can be conceptualized as a statistical distribution of partially bundled allelic pathways of descent each characterized by its own unique coalescent pattern. The many distinctions yet connections between notions of phylogeny at the levels of genes versus populations have made phylogeography a rich point of contact between the traditionally distinct fields of population genetics and phylogenetic biology.

Science often is serendipitous, as the following stories well illustrate. Shortly after joining the University of Georgia as an Assistant Professor in 1975, I gave a departmental seminar describing work on allozyme variation in fishes. Echoing a sentiment popular at the time, I concluded that regulatory rather than structural genes should be studied next because changes in gene regulation were perhaps at the heart of adaptive evolution. I queried the audience for suggestions on how I might examine regulatory genes, and one responder asked whether I had considered using restriction enzymes to assay repetitive nuclear DNA sequences, which at the time were viewed as prime candidates as regulatory modulators. I had never heard of restriction enzymes! However, the idea was intriguing so I soon approached several faculty members at the University in an attempt to identify a collaborating laboratory where I might learn restriction digestion techniques. To my chagrin, the inquiries met with cool responses, except one: Dr. Robert Lansman welcomed me to his laboratory, but noted with apology that he had limited experience with nuclear DNA and instead conducted research on the biochemistry and cellular biology of mitochondrial DNA. I barely had heard of mitochondrial DNA! However, left with few options, I accepted Bob's offer to gain familiarity with DNA-level assays.

Before long, we were generating agarose gels with mtDNA restriction profiles, initially from small mammals. Although I was still viewing the effort mainly as a training exercise, intriguing questions began to emerge. Why did each individual display only a few mtDNA bands on a gel, rather than a smear of fragments from the billions of mtDNA molecules that must be included in an assay? (It must be because each specimen had a specifiable mtDNA genotype with respect to the restriction sites assayed.) Why did different mice within local populations often display distinct restriction fragment length polymorphism (RFLP) patterns, such that observed mtDNA variation primarily was distributed among rather than within individuals? (With hindsight, it must be because mtDNA mutations arise frequently, and sometimes precipitate within a small number of animal generations a genotypic turnover in the population of mtDNAs in a germ-cell lineage from which the assayed soma were derived.) Why did mtDNA genotypes in organismal populations appear connectable to one another in phylogenetically intelligible ways? (Because intermolecular recombination must be rare or nonexistent in these maternally inherited molecules, such that the matrilineal histories of mutation events were recorded in extant mtDNA genotypes.) Why did members of sexually reproducing species usually group together by mtDNA genotypes when the evolutionary connecting agents of mating and genetic recombination seemed not to apply to these asexually transmitted genomes? (Because, as we now know, coalescent processes ensure phylogenetic links among genotypes via vertical pathways of ancestry even in the absence of interlineage genetic exchange mediated by mating events.) What ramifications might stem from the heretical practice made possible by mtDNA of viewing haplotypes as clones and individual animals as operational taxonomic units (OTUs), in population genetic analyses? (The list of responses is now long.)

In general, many unorthodox perspectives on evolution eventually were to emerge from studies of mtDNA, but years would pass before relatively clear answers to some of the questions listed above and others similar to them were to be forthcoming. The lag time reflected in part the difficulty experienced by many researchers (certainly by me) in reorienting thought away from the traditional Mendelian perspectives that applied so well, for example, to allozyme systems on which many of us had been trained.

My collaboration with the Lansman laboratory went well, and our first paper on mtDNA variation in a natural population soon appeared (in 1979), followed shortly thereafter by the first large-scale phylogeographic survey of any species based on mtDNA lineages (also in 1979). The technical stage for these efforts had been set in the early 1970s through prior mtDNA research on several fronts. For example, Brown & Vinograd in 1974 and Upholt & Dawid in 1977 had demonstrated the feasibility of generating restriction enzyme cleavage maps for animal mtDNAs; Dawid & Blackler in 1972 and Hutchinson and colleagues (among others) in 1974 had documented predominant maternal inheritance for mtDNA in higher animals; and Upholt in 1977 had developed a statistical procedure for estimating sequence divergence among mtDNA genotypes from comparisons of restriction digests. Furthermore, in the same year that our first phylogeographic works appeared in print, Brown and colleagues published an influential article highlighting the unexpected fast pace of mtDNA sequence evolution as gauged by interspecies comparisons of higher primates.

In the late 1970s, excitement generated by the new mtDNA discoveries ran high. I remember pondering the many research possibilities, of which two of anecdotal interest can be mentioned. Early on, it occurred to me that mtDNA might be a wonderful tool for analyzing the evolution of parthenogenetic vertebrates, for at least two reasons. First, all such unisexual biotypes were thought to have arisen through hybridization between sexual species, such that by utilizing mtDNA data it should be possible to identify the maternal parent taxon in each case. Second, because parthenogenetic taxa reproduce asexually, the history of maternal lineages within them should in principle be one-and-the-same as the entire organismal phylogeny (unlike the case for a sexual species). I remember reasoning that it would be safe to shelve these ideas for the moment in the belief that many years would elapse before any molecular biologists might dream of this obscure biological application for mtDNA. I could not have been more wrong. One of the first mtDNA analyses of natural populations dealt with precisely these evolutionary issues in parthenogenetic lizards! Eventually, my laboratory did examine evolutionary processes in gynogenetic and hybridogenetic fish using mtDNA, but only well after Wes Brown, Craig Moritz (Brown's postdoctoral researcher at the time), and their associates had produced an important series of mtDNA papers on the origins and evolution of parthenogenetic reptiles.

It also seemed evident to Bob Lansman and myself that mtDNA analyses of human populations would be of great interest. However, we elected not to pursue this topic. Personally, I was wary of the inevitable social and political fallout from whatever findings might be uncovered about the nature of genetic differences between human skin color races, or between humans and great apes; and, in any event, it seemed likely that the necessary research would be accomplished by someone. I also reasoned that I was more suited for non-human natural history studies. An influential study on human mtDNA evolution did famously appear in 1980, followed later by many more mtDNA analyses of higher primate phylogeny and human geographical variation.

I should digress from this personal account for a moment to relate the history of Wes Brown's involvement with mtDNA, because this traces the other major root of evolutionary interest in the molecule. The story began in 1968 when Brown went to Caltech as a graduate student and was introduced to mtDNA in the laboratories of Giuseppi Attardi and Jerome Vinograd, where mtDNA transcription and physical chemistry, respectively, were being studied. In 1971, Brown went to an exhibition of Max Escher paintings at the Los Angeles County Museum, where he happened to meet John Wright, the curator of the herpetology department. Wright was probably the most knowledgeable person in the world on *Cnemidophorus* lizards, and Brown's chance meeting with him that day was to lead to their collaborative studies on the evolutionary origins of parthenogenetic taxa from a genealogical perspective. Brown gathered mtDNA data at Caltech from 1971 to 1973 but, as mentioned above, the first papers did not appear until several years later. After a postdoctoral stint at the University of California at San Francisco, Brown moved across the Bay in 1978 to join Allan Wilson's group at Berkeley. There he restructured and equipped the laboratory for studies of animal mtDNA, and among other efforts initiated the important research mentioned above on human genealogical evolution.

Returning to the developing story at the University of Georgia, in those early years another important event for me personally stemmed from a casual conversation over lunch. I was explaining to my colleague Michael Clegg our recent findings on modes of inheritance and patterns of geographical variation in mtDNA for small mammals, and he mentioned that the issues seemed analogous to those for surname evolution in many human societies. This simple comment struck home, and helped greatly in my otherwise tortuous transition from Mendelian to phylogenetic thinking at the intraspecific level. The surname analogy does indeed hold well. Just as sons and daughters "inherit" their father's nonrecombined surname (before recent rule changes in some societies), so too do progeny normally receive nonrecombined mtDNA from their mothers. Furthermore, much the way that mutations sometimes arise in surnames (my own name was a nineteenth century misspelling of "Avis"), point mutations occasionally arise and cumulatively differentiate related mtDNA genotypes. Thus, mtDNA molecules record matrilineal histories much as surnames record patrilines, except that the matrilineal records extend much further back in time (surnames were invented de novo only within the past few centuries).

These insights were new to me, but not completely so to the field. Beginning as early as 1931, statistical demographers had studied the dynamics of surname turnover in human populations, using models that now could be applied often with little modification to gene lineages such as those provided by mtDNA. Such models stimulated my own and my students' efforts to examine the theoretical ties between population demography and phylogeographic patterns within and among populations and species, and to address these expectations in a series of empirical mtDNA studies on a wide variety of organisms in nature. "Coalescent theory" is the term now applied to the formal mathematical and statistical properties of gene genealogies, and the results from this discipline are highly relevant to molecular phylogeographic interpretations.

In addition to these and several other signal events, throughout the 1980s and 1990s there was a burgeoning growth in the application of both genealogical theory and molecular data to phylogeographic analyses. This included extensions and refinements of coalescent theory for populations of varying demographies, improvements in statistical and cladistic procedures for extracting phylogeographic information from empirical data on gene genealogies, and a great plethora of empirical applications primarily involving mtDNA. Of course, progress in several related areas, not the least of which are molecular and computer technologies, also contributed significantly to the general scientific climate that permitted the flowering of phylogeographic studies during the last few decades.

Phylogeography as a recognizable discipline is historically recent, having grown from empirical molecular analyses of mtDNA in conjunction with mathematical studies of coalescent processes that are necessary to capitalize on this new class of genealogical information within species. The field has had an auspicious start, but the greatest benefits and opportunities for phylogeography will continue to arise, as they have in the past, from the field's central, integrative, position within the evolutionary and ecological sciences.

11

Cladists in Wonderland

At least in the last half of the twentieth century, nothing quite paralleled the "cladistic revolution" in terms of mixing (inappropriately) ideology with science. Following the publication in 1966 of *Phylogenetic Systematics,* an English translation of Willi Hennig's 1950 treatise originally in German, at least two academic generations of systematists were often caught up in a religious-like fervor, trumpeting the virtues of cladism and crusading to convert nonbelievers and agnostics to their scientific faith. Hennig's phylogenetic principles were indeed important if not revolutionary in biologists' attempts to understand the various sources of evolutionary similarities and differences among organisms, but what seems even more amazing is how this one particular evolutionary topic generated so much ideological fervor. Although many important scientific contributions have emerged from cladistic reasoning, cladists often were led too far by their dogmas. Perhaps nowhere was this more true that with regard to species concepts. The following is Avise's review of one edited book published under the cladistic banner. As you will see, he could find no better words to describe some of the cladist's views than those issued by fictitious characters in Lewis Carroll's *Through the Looking Glass* and *Alice's Adventures in Wonderland.*

"A hill can't be a valley, you know. That would be nonsense," said Alice. The Red Queen shook her head. "You may call it 'nonsense' if you like, but I've heard nonsense, compared with which that would be as sensible as a dictionary."

In the year 2000 I attended a conservation genetics symposium at which one of the speakers claimed to have caught a speciation event in the act. Earlier in this century, tiger beetles (*Cicindela dorsalis*) were distributed more or less continuously along the eastern coast of the United States, but shoreline development extirpated populations in the mid-Atlantic states, causing a range disjunction between still-extant populations in New England and the southern states. By analyzing DNA sequences in living and museum-preserved specimens, small but detectable nucleotide differences were uncovered between these two extant populations, a distinction not formerly possible because mid-Atlantic populations had been polymorphic for the sequences in question. Under one version of the phylogenetic species concept (PSC), a new species had arisen precisely when the polymorphism became a fixed difference (in this case, via the extinction of intermediate demes). If this speaker's PSC-based conclusion about species formation is to be taken seriously, and generalized, conservation biologists might naively rejoice. In the coming decades, as natural populations of many species are extirpated or reduced to small inbred units, intraspecific polymorphisms increasingly will be converted to fixed allele differences between allopatric demes. Under PSC logic, by definition, this will result in a great proliferation of new species. Thus, we may look forward to a twenty-first century in which the rate of species origin (via fixation of genetic variants) may far out-pace the rate at which currently recognized taxonomic species are driven to extinction. What most biologists had feared as a deepening valley in species numbers may instead soon become a numerical peak in taxonomic species richness!

"It was much pleasanter at home," thought poor Alice. "I almost wish I hadn't gone down that rabbit-hole—and yet—it's rather curious, you know."

To scientists raised under the traditional Biological Species Concept (BSC) of Theodosius Dobzhansky and Ernst Mayr, where the evolution of intrinsic (genetically based) reproductive barriers is the underlying basis of cladogenesis, the speciational world of various cladistic camps can seem as curious as that encountered by Alice in her sojourns down a rabbit's hole or through a looking glass. To us outsiders, it can be a world of sense and nonsense often turned on its head, of erudite jabberwocky, of impeccably logical illogic, of surreal reality. Now, for the first time, this speciational wonderland is fully explored in a single volume. In *Species Concepts and Phylogenetic Theory*, Wheeler and Meier have assembled a collection of invited articles eloquently portraying a conceptual world that is simultaneously as coherent and incoherent as anything conjured up by Lewis Carroll. I heartily recommend this entertaining treatise to anyone interested in some recent developments (not advances) in phylogenetic reasoning as applied to species issues.

An opening chapter by Joel Cracraft wisely enjoins readers (p. 6) "to grab a favorite fetish and conjure up a bit of luck" in interpreting what will follow. The rest of the book is arranged in an illuminating debate format. In the first section, leading proponents of various phylogenetic, Hennigian, and evolutionary species concepts clearly lay out their respective platforms on the "species problem." This section is followed by counterpoint chapters, written by the same sets of authors, as critical responses to the position statements of their adversaries. The closing chapters are composed of formal rebuttals to these counterpoints. The only unfortunate aspect of this format stems from inevitable delays in the sequential give-and-take of written exchanges: The book was long in production, and little relevant literature (other than that by the authors themselves) is cited post-1993.

"Off with his head!"

Ernst Mayr was the primary traditionalist from Alice's native world to accept an invitation to this phylogenetically oriented tea party. Predictably, he and the BSC are guillotined again for hackneyed reasons such as: The BSC is based on population criteria (intrinsic reproductive isolation [or its converse, potential to interbreed]) that are often difficult or impossible to measure directly in nature, especially among allopatric forms; it can lead to gray areas in species assignment when genetic isolation is incomplete; interbreeding is a plesiomorphic rather than apomorphic condition, so it cannot be a valid basis for identifying Hennigian clades; the BSC does not apply to asexual organisms and, thus, is admittedly nonuniversal; BSC guidelines are inexplicit on precisely where to draw species boundaries within a temporal anagenetic sequence of ancestor-descendent populations; and the BSC carries needless baggage because it references a causal evolutionary process (when instead a species should be identifiable by an idea-free definitional algorithm).

Mayr articulately responds (yet again) to such charges, with an evident sense of conviction that Alice's home world is the one that is sane, rather than the world of the March Hare and Dormouse. Yet at times he seems thoroughly exasperated by the tea-party repartee (p. 93): "For someone who has published books and papers on the biological species for more than 50 years, . . . the reading of some recent papers on species has been a rather troubling experience;" and (p. 163), "I realize that it is apparently distasteful for a cladist to read anything not written by another cladist."

> *Tweedledum and Tweedledee agreed to have a battle . . . "I know what*
> *you're thinking about," said Tweedledum, "but it isn't so, nohow."*
> *"Contrariwise," continued Tweedledee, "if it was so, it might be; and*
> *if it were so, it would be; but as it isn't, it ain't. That's logic."*

The more novel aspects of this book are the internecine debates clearly portrayed among various phylogenetically oriented camps. At least four distinct definitions of species exist in this alternative world, and their respective advocates effectively challenge one another's notions as well. Given all the hoopla in the recent literature about the purported demise of the BSC, one might suppose that the revolutionaries had come up with something far better. This volume indicates that they have not. Ironically, this failure to obtain a compelling synthesis on cladistic species notions may be the book's primary (but unintended) contribution. What follows are these four alternative species definitions and some brief hints at their serious shortcomings (often exposed by the authors themselves).

The shop seemed to be full of all manner of curious things—but the oddest part of it all was that, whenever Alice looked hard at any shelf, to make out exactly what it had on it, that particular shelf was always quite empty.

Wiley and Mayden define a species as "an entity composed of organisms that maintains its identity from other such entities through time and space and that has its own independent evolutionary fate and historical tendencies." In an effort to clarify, the authors add (p. 75): "To say that an evolutionary species has its own evolutionary fate is simply to say that it is a real entity and not a figment of our imagination."

The Wiley–Mayden definition of species is an attempt to resurrect George Gaylord Simpson's evolutionary species concept (ESC), which Simpson himself mostly abandoned in 1961 as being rather nebulous for systematic purposes. Although the ESC is certainly compatible with the "modern evolutionary synthesis" (and indeed can be interpreted, in effect, as the BSC extended through time), other authors in the current volume likewise see major problems in deriving operational procedures from ESC definitions. Meier and Willmann view the concept as entirely subjective because (p. 178) it "fails to provide criteria that allow a nonarbitrary delimitation of species in both the temporal and spatial dimensions." Mishler and Theriot find the ESC to be (p. 129) "incoherent to us because it fits everything and nothing." Wheeler and Platnick dislike it because (p. 142): "Saying that lineages exist or that they have histories or tendencies or rates are no more than vague assertions that evolution exists." Mayr asks (p. 97) "What population in nature can ever be classified by its historical fate when this is entirely in the future?"

"I think you might do something better with the time," Alice said, "than wasting it in asking riddles that have no answers."

Next, Meier and Willmann provide a species definition, which they term the "Hennigian species concept," that at first glance looks much like the BSC: "Species are reproductively isolated natural populations or groups of natural populations. They originate via the dissolution of the stem species in a speciation event and cease to exist either through extinction or speciation." One point of departure from the BSC is the authors' notion that a "stem species cannot survive speciation," an artificial convention necessitated by their operational paradigm that "a species must comprise the entire branch segment between two speciation events." But why, apart from definitional cleanliness, must a stem species instantly dissolve or go extinct? Mayr wonders, for example, why a large ancestral population that budded off a reproductive isolate (e.g., by a founder event at the periphery of its range) must also be viewed suddenly as a new species (p. 94): "Because the 'new' species is evidently the same genetically as the old species, I do not understand how it can be called new."

Another point of departure from the traditional BSC is Meier and Willmann's insistence on a *complete* absence of gene flow for the species status of contemporary taxa (p. 40): "Absolute isolation requires that even if hybrids between species occur, these hybrids are not able to successfully backcross with members of the parental population;" and (p. 41), "Only when absolute isolation is used as the sole species criterion are objective and mutually comparable units delimited." However, as noted by Wiley and Mayden (p. 157), "we know of no recently evolved and closely related species of North American freshwater fish that is 100% reproductively isolated from its sister species," and (p. 205), "we are aware of gene introgression among distantly related (non-sister) species" also.

Similarly, Mishler and Theriot note (p. 179) that "nearly absolute isolation may exist between groups at some extreme levels of divergence, but in many plants that would be at the level currently ranked at about the family level." Thus, both sets of critics agree that if Meier and Willmann's requirement of absolute isolation were used as the ranking criterion, orders of magnitude *fewer* species would be recognized than now. This is the opposite to the anticipated ballooning in species numbers that would accompany the next two species concepts discussed in this volume.

"Curiouser and curiouser!" cried Alice.

The other two species concepts debated in the book are alternative versions of the PSC. Mishler and Theriot define a phylogenetic species as "the least inclusive taxon recognized in a formal phylogenetic classification . . . Taxa are ranked as species rather than at some higher level because they are the smallest monophyletic groups deemed worthy of formal recognition . . ." To Wheeler and Platnick, a phylogenetic species is "the smallest aggregation of (sexual) populations or (asexual) lineages diagnosable by a unique combination of character states." Both versions of the PSC equate "species" with monophyletic units, one difference being that the Wheeler–Platnick version would name all such units (no matter how small) as taxonomic species, whereas the Mishler–Theriot version would reserve formal taxonomic recognition to those units deemed especially worthy by virtue of the "amount of support for their monophyly and/or because of their importance in biological processes operating on the lineage in question." This too seems remarkably vague.

Both versions of the PSC focus on issues of diagnosability (based directly on character evidence, notably synapomorphies) far more than on any underlying formational evolutionary processes. Wheeler and Platnick are most explicit about this (p. 59): "Speciation is marked by character transformation . . . The moment of speciation is, in theory, precise and corresponds to the death of the last individual that maintained polymorphism within a population" (as, for example, in the tiger beetle case study that opened this review). By equating speciation with character transformation, Wheeler and Platnick (p. 61) believe that they are divorcing the analysis of phylogenetic patterns from unnecessary assumptions about evolutionary process, and are thereby transforming speciation analysis into a rigorous science. Furthermore, in their fanaticism against pheneticism in species concepts, Wheeler and Platnick are forced to disregard altogether any information on the amount of genetic divergence (or even synapomorph numbers) in defining terminal "clades" that they wish to call species.

To at least some molecular population geneticists trained in Alice's native world, this myopic focus on species diagnosability by even minute differences may seem utterly nonsensical. In sexual organisms, each individual typically carries a unique combination of genetic variants due to recombinational shuffling of existing Mendelian polymorphisms. Furthermore, thousands of de novo mutations arise and spread relentlessly in populations. Given limited organismal dispersal (and sufficient resolution in the molecular assays), one or more of these synapomorphs often will differentiate regional populations, local demes, extended kin groups, and even nuclear family units (not to mention individuals and sometimes subsets of their constituent somatic cells!). Do we really wish to recognize every such diagnosable unit as a distinct species simply because its members happen to share one or a few derived genetic mutations that we may have detected in our available assays?

Even some of the cladists see a problem here. Willmann and Meier write (p. 105): "Potential users of Wheeler and Platnick's phylogenetic species concept should be aware that, according to their species definition, the number of character fixations is equal to the number of species, because with every change a new combination of character states is produced." Thus, (p. 103) "even a mutant strain of *Drosophila melanogaster* in a culture vial would constitute a separate phylogenetic species if all members carry the mutation." Extending this concern, Meier and Willmann consider a hypothetical herd of deer housed in the basement of the American Museum of Natural History (p. 173): "Given that the deer . . . are a family group, they probably have some unique genetic markers and would probably constitute a phylogenetic species sensu Wheeler and Platnick."

Wheeler and Platnick reply (p. 60), "It has been suggested that our phylogenetic species concept is problematic because it may result in an enormous number of species . . . Our response is, so what?" This blithe attitude is inadequate. It may be one thing to advocate (as Wheeler and Platnick do) an epistemological avoidance of underlying

concepts in attempting to describe evolutionary processes from observable patterns, but quite another to completely disregard what we
do know about underlying ontological reality (in this case regarding
such evolutionary processes as mutation, sexual Mendelian inheritance, and genetic recombination) in developing our biological classifications. At least the Mishler–Theriot version of the PSC seems
clearer about such issues by reserving species assignments for the
somehow more salient of the terminal monophyletic units in the biological world. Still, Mishler and Theriot write (pp. 45 and 46) that
"breeding criteria in particular have no business being used for group-
ing purposes" and "apomorphies [are] considered to be the necessary
empirical evidence for unambiguous phylogenetic species."

When all is said and done, much of the brouhaha between the PSC
and the BSC often boils down to what taxonomists might wish to do
with what now are considered geographic populations. After all, if
two or more groups of sexual organisms coexist sympatrically without interbreeding, by definition they are good biological species. For
this same reason (and only for this reason, although many cladists
seem loath to admit it), they would also be valid phylogenetic species.
The remaining practical question is what to do with allopatric populations. Wheeler and Platnick would formally recognize a phylogenetic
species as any current population distinguishable from others by so
little as one derived mutation. By contrast, most followers of the BSC
would wish to rank as subspecies (using a Latin trinomial) only the
better-marked of the geographic populations within a "polytypic"
species. As stated by Mayr (p. 26), "the use of trinomials conveys two
important pieces of information: closest relationship and allopatry.
Such information is valuable, particularly in large genera."

Granted, we humans can choose to make words mean what we will,
and in this sense, there can be no ultimate right or wrong in species
definitions. Those with entrenched views (on any side of the debate)
can always survive by digging in their definitional heels. But to remove (as would may cladists) all reference to evolutionary genetic

processes (including reproductive relationships) in species concepts seems to me (and Mayr) a big step backward to the old days of typological thought.

> *"When I use a word," Humpty Dumpty said in a rather scornful tone, "it means just what I choose it to mean—neither more nor less." "The question is," said Alice, "whether you can make words mean so many different things." "The question is," said Humpty Dumpty, "which is to be master— that's all."*

At the outset, Cracraft warns the reader: "The literature on species concepts is riddled with confusion and obfuscation" (p. 7); "arguments and conclusions using the same words might not mean the same thing" (p. 5); and "definitions do not necessarily make things real" (p. 11). Indeed, the speciational wonderland revealed in this volume is rife with subterfuge and shifting meanings. For example, in their zeal to remove all explicit mention of interbreeding and reproductive relationships in species definitions, the cladists have fully adopted Hennig's (1966) term "tokogeny," which Meier, Willmann, Wheeler, and Platnick nonetheless use virtually synonymously with interbreeding. However, according to Mishler and Theriot, tokogeny is (p. 127) "the diachronic relationship through time between a parent and an offspring" (and thus applies to asexual as well as sexual taxa). Overall, much of the species debate in the cladistic world hangs on such (re)inventional word games.

In another such example, Wheeler and Platnick redefine "character" and "trait" in an unconventional, stricter sense (p. 196): "A trait is an attribute that varies among individuals or populations within a single species . . . A character is an attribute that varies between but not within species." This definitional sleight of hand may be a useful device for escaping some of the pitfalls of equating every change in a (conventional) trait with a speciation event, but it also introduces considerable lexical confusion (not to mention circular reasoning) to the broader debate.

Said the Cat: "we're all mad here. I'm mad. You're mad." "How do you know I'm mad?" said Alice. "You must be," said the Cat, "or you wouldn't have come here."

In various passages throughout Wheeler and Meier's volume, readers from Alice's original realm may well begin to question the very notion of sanity itself in the cladist world. So too do some of this book's authors themselves, as for example when the cladistic "fear of paraphyly" is parodied by Wiley and Mayden (p. 82): "We have a hard time accepting a 'thing' as paraphyletic unless lots of other 'things' are paraphyletic. Consider Ed Wiley. From one point of view, he is an individual. From another point of view, he is a group of cells. Why not apply these terms to Ed Wiley? He is, after all, a kind of a group. Ed Wiley has three children. They do not reside in his body, nor are they all named 'Ed Wiley'. Obviously, following [such] reasoning, Ed Wiley is paraphyletic. He might even be considered polyphyletic if one followed Nelson (1971). If Gary Nelson wished to apply the term *paraphyly* to individual organisms, he might assert that Ed Wiley does not exist . . . This leaves Aaron Wiley in a fix . . . Yet Aaron exists. De Queiroz and Donoghue (1990) would accept the paraphyletic Ed Wiley; they would just claim that his primary spermatocytes are actually more closely related to Aaron than to Ed's own brain cells. Mishler and Brandon would deny that Ed and Karen exist or would assert that they are only collections of cells. Aaron exists only until such time as he has children."

So Alice sat on, with closed eyes, and half believed herself in wonderland, though she knew she had but to open them again and all would change to dull reality.

After reading *Species Concepts and Phylogenetic Theory*, readers may fairly ask, "Is this all there is to the crowning legacy of nearly 20 years of cladistic hyperbole on species concepts? Is this really the stuff to which the BSC must be abandoned? Was the conventional reality on biological species really so dull, so grotesquely inadequate, so positively misleading that an entirely new wonderland of species concepts was necessitated?"

This book is indeed an illuminating and valid compilation of thought about speciation from the realms of Hennigian as well as transformed cladistics. Thus, it succeeds as a historical treatment and as an extended case study in the sociology of science. It also reveals more clearly than any BSC proponent ever could that most of the recent cladistic foundational pillars really are insecure when it comes to species concepts.

Yet, there are bona fide advances in phylogenetic aspects of speciation theory, spurred at least in part by a response to this cladistic revolution, that are totally missing from this book. I'm referring to a large and growing body of relevant phylogenetic thought and literature coming not from the cladistic world per se, but rather from the more Aliceian realms of molecular biology, population genetics, and coalescent theory.

> *"Why, it's got branches, I declare! How very odd to find trees growing here!" said Alice.*

When studies of maternally inherited mitochondrial DNA (mtDNA) were introduced to population biology in the late 1970s, they soon led to the revolutionary concept that asexual, nonanastomose, hierarchical gene genealogies exist even within what are otherwise sexually reproducing populations. This motivated the now conventional distinction between gene trees (of which great numbers occur within any extended population pedigree) and population trees or species phylogenies. This perspective carries a host of ramifications for species concepts, not the least of which is the recognition that it makes little biological sense to focus unduly on single diagnostic genetic characters, including synapomorphies, as a basis for distinguishing sexual species (in part because, from first principles of population genetics, the historical transmission pathways of alleles vary from gene to unlinked gene, and often will be inconsistent in the molecular clades they describe). It also helped spawn the rise of modern coalescent theory, which addresses how the historical demographies of populations impact (indeed, are virtually inseparable from) genealogical patterns.

Nowadays, *phylogenetic aspects* of biological speciation processes should center on the following sorts of questions: Exactly how do various nonequilibrium population dynamics (and natural selection) influence gene-tree structures? What are the ramifications of appreciating that a traditional sticklike cladogram for sexually reproducing taxa is really a statistical "cloudogram" of gene trees with a variance? How many genealogical pathways are needed to estimate major disjunctions in an organismal phylogeny that we might wish to formally name, or perhaps taxonomically earmark for special conservation efforts? How do various kinds of genealogical concordances and discordances arise among multiple gene trees within an extended organismal pedigree, and what are their relevances to salient biological discontinuities at the population or "species" level? If speciation is to be viewed properly as an extended temporal process (rather than a point event as in the oft black-and-white cladistic world), then what are the means and variances in the temporal durations of this biological phenomenon?

Under modern phylogeographic perspectives on species, there is no inherent conflict between the criterion of historic reproductive isolation and *properly formulated* phylogenetic criteria in accounting for salient biological discontinuities in nature that we might wish to call a species. In this phylogeographic approach, scientists begin with basic Mendelian, population-demographic, and population-genetic principles, toss in a large dose of historical geographic considerations, and thereby produce a synthetic conceptual framework for species recognition that attempts to fully integrate the better elements of the traditional BSC and PSC.

Alice laughed. "There's no use trying," she said: "one can't believe impossible things." "I daresay you haven't had much practice," said the Queen.

In this attempted synthesis based on phylogeographic and coalescent principles, both reproductive and phylogenetic criteria are seen as intimately related concepts, and as mutually informative aspects of what is usually a temporally extended speciation process for sexual organisms. Reproductive barriers are important for species concepts (even within a strict phylogenetic framework) because, through time, they generate and promote increased genealogical depth and concordance across composite DNA-transmission pathways. Conversely, phylogenetic considerations are important (even within the philosophical framework of the BSC) because they force explicit attention on historical and demographic aspects of the speciation process. Furthermore, an explicit focus on population demographic aspects of historical lineage sorting may also go a long way toward explaining how and why biological discontinuities often appear to exist among asexually reproducing organisms.

Perhaps the ongoing phylogeographic synthesis that tries to wed (rather than divorce) phylogenetic and reproductive concepts in species recognition will prove to be only yet another fantasy. But I doubt it. Instead, I have great hope that the peculiar tea-party banter between the Aliceians and the Mad Hatters over species concepts will eventually clarify, and that a more intelligent dialogue and eventual synthesis will emerge. If so, the 20-year quarrel between proponents of the BSC and the PSC, so cogently encapsulated in the Wheeler and Meier volume, will someday be remembered as little more than a "tempest in a teapot."

12

Evolving Genomic Metaphors: A New Look at the Language of DNA

> Metaphors in science can be powerful as well as fun, and Avise has been adept in their use. He here introduces a new set of metaphors for describing some recent and unanticipated discoveries about the structures of nuclear genomes, especially as revealed in the Human Genome sequencing project whose first phase was completed in 2001. (For another extended metaphor on genetics and evolution, readers might wish to read the author's book: *The Genetic Gods: Evolution and Belief in Human Affairs,* 1998.)

Metaphors in science are like foghorns and lighthouses: They usually reside in treacherous areas, yet they can also guide research mariners to novel ports. With the recent flood of DNA sequences from the human gene pool and those of other eukaryotic species, the exploratory ship of biology is suddenly up to its gunnels in molecular data portraying a genomic seascape that is far more turbulent and evolutionarily fluid than formerly envisioned.

Evocative metaphors can distill an ocean of information, whet the imagination, and suggest promising channels for navigating uncharted genetic waters. For example, the metaphor of nucleotide sequences as encrypted language, translatable to the plain text of polypeptides, may have facilitated research in the 1960s that cracked the "genetic code." In a more recent example, the notion of the genome as a "book of life" helped to focus and sell the human genome sequencing project. However, metaphors can also mislead. The metaphor of the genome as a well-crafted blueprint or a finely tuned machine may have blinded many biologists to genomic imperfections attributable to phylogenetic constraints and evolutionary-genetic trade-offs.

Clearly, metaphors vary in utility and can influence research programs.

In the twentieth century, "beads-on-a-string" was a prevailing metaphor for how housekeeping genes (those that encode proteins) were densely packed along each chromosome. Draft sequences of the human genome have nailed the coffin shut on that caricature: the coding "beads" make up less than 2% of our DNA, and most are themselves subdivided into beadlets (exons) interspersed with noncoding introns that comprise more than 95% of a typical transcription unit. Accordingly, some scientists next visualized protein-coding genes as tiny scattered oases in a genomic desert, implying that all else was a wasteland. Fortunately, this view did not prevent the genomic outback from being reconnoitered in the human genome project, because the results were truly incredible.

The intergenic wilderness proved to be populated by a motley crew of intriguing genetic characters: active promoters and regulators of gene expression, comatose pseudogenes, descendants of immigrant DNAs (perhaps horizontally transferred from microbes), vagabond sequences, hordes of tandem short repeats, and great armies of repetitive elements—some with thousands of like-uniformed members. Astonishingly, at least 50% of the human sequence is derived from transposable elements (TEs) that have dispersed themselves across the genome either as mobile DNA or via reverse-transcribed RNAs. Some of these smaller jumping genes are freeloaders that hitch rides on the backs of larger roving elements, like mites on fleas.

Nonetheless, the earlier metaphor of the intergenic region as barren desert probably still acts to divert attention from what could be highly fertile research terrain. Ironically, this genetic hinterland of regulatory tacticians, renegades, deadbeats, ramblers, and foreigners may be the real mother lode for deep intellectual treasures regarding life's functional and evolutionary modes. By prospecting and mining rich research veins for the interactions between protein-specifying genes and the great assortment of repetitive elements, regulatory sequences, and other noncoding DNAs, geneticists only lately have begun to excavate precious conceptual ores and jewels from these genomic quarries.

A long-standing genomic metaphor has described all genetic material as selfish, each DNA segment (functional or not) concerned first and foremost with its own transmission. One effective strategy for a DNA element is to contribute to the health of its host, because such behavior raises the element's hereditary prospects. However, the strategies of other "selfish genes" may harm the individual. Hence arose the metaphor of "parasitic DNA," which posits that many genetic elements reside and replicate within the genome at organismal expense.

By proliferating across the genome, mobile elements promote their own survival, but their sheer numbers probably add a metabolic burden to the cell. Furthermore, these genetic nomads are a major source of mutations, most of which are deleterious to the host. Is the parliament of good-citizen genes powerless during the evolutionary process to constrain these genomic outlaws? No, because natural selection at the level of organismal fitness in effect polices the net product of all DNA-level interactions, and is the final arbiter in all matters of genetic jurisprudence.

The genomic encyclopedias of life are revealing many surprising ways that transposable elements and housekeeping genes have coevolved within their cramped cellular quarters. Most important is the realization that some TEs (or their immobile offspring) also confer significant benefits to host genomes. For example, many TEs carry regulatory sequences that over evolutionary time have been drafted into the adaptive service of modulating gene expression. Many salubrious tasks for TEs have likewise been "host recruited," such as sponsoring variation at histocompatibility loci by serving as recombination templates, forming centromeric regions, replenishing telomeres, and promoting mutations and gene duplications that provide a fodder for evolutionary innovation.

Geneticists are gradually abandoning the view of intergenic regions as mere junk. If this metaphor is retained at all, it should be modified to picture these genomic tracts in the way that many anthropologists now view ancient garbage dumps—not as containing rubbish, but as holding important clues to people's daily lives in civilizations past. Likewise, DNA sequences outside the exons may be uniquely revealing about the coevolutionary lives of DNA within cell lineages. In short, metaphors can and should evolve to accommodate new findings.

One adaptable metaphor would liken each genome to a social collective whose DNA sequences display intricate divisions of labor and functional collaborations, yet that maintain partial autonomies of fate (due to sexual reproduction), resulting in occasional conflicts of interest. In this view, many types of DNA behavior roughly mirror those of humans bound in tight social arrangements, such as communes. These include not only collaborative efforts, but also cheating; aggregate actions, but also personal opportunism; group alliances, but also conflicts; and parliamentary needs often opposed by egoistic tendencies. However, this metaphor neglects the clonal proliferation of elements possible within a genome, and the extensive reassortment of unlinked sequences that accompanies each generation of sexual reproduction. An Israeli kibbutz might be a closer analogue of genomic society in this latter regard, because most marriages are outside the collective. It would be an even better analogue if a partially randomized clique of kibbutznikim in each generation married a comparable suite from another conclave to initiate each new commune!

Another metaphor might present each genome as a miniaturized cellular ecosystem with each gene occupying a particular functional niche, yet in which the DNA sequences have evolved elaborate interactions (including parasitism, commensalism, and mutualism) normally associated with species in natural biological communities. This metaphor falls short, however, by failing to ascribe to genes the exceptional collaborative responsibilities also entailed in producing a discrete entity (the organism) whose survival and reproduction is key to the evolutionary game.

The hope for any metaphor in science is that it may bring otherwise unfamiliar subjects to life, make connections not otherwise apparent, and stimulate fruitful inquiry. A danger is that a metaphor can restrict rather than expand research horizons. Many genomic metaphors have elements of truth, and each may have its time and place. I doubt, for example, that a depiction of the genome as a molecular ecosystem would have served well in promoting or guiding the human genome project. However, perhaps the time is right for new panoramic images of the genomic landscape that capture proper notions of complexity and evolutionary dynamism. Although no one metaphor is likely to be informative in all respects, new perspectives that view the genome as an interactive community of evolving loci may be especially useful and stimulating at this time.

13

Genetic Mating Systems and Reproductive Natural Histories of Fishes

Avise has taught ornithology throughout his career (among many other courses), and thus was well aware that molecular markers had revolutionized thought about avian mating systems. In particular, they challenged conventional wisdom that most songbirds are strictly monogamous. Markers revealed that foster progeny (resulting from stolen fertilizations or egg dumping) occur in the nests of many avian species. In the mid-1990s, much of the Avise lab converted from phylogeographic studies to genetic assessments of parentage and mating systems in many kinds of creatures, especially fish. At that time, fishes were a near-virgin taxonomic group for genetic research on reproductive behaviors in nature, almost completely ignored by molecular ecologists despite the existence of a rich natural history literature. The great diversity of procreative modes in fishes puts those of birds (for example) to shame, as this chapter reviewing relevant genetic discoveries attests. (For more on nature's peculiarities genetically revealed, see the author's *Genetics in the Wild* [2002].)

Fish have remarkably diverse reproductive behaviors. A rich natural-history literature documents mating systems ranging from pelagic group spawning to cooperative breeding to social monogamy. Subsequent to spawning, adult care of fertilized eggs and larvae may be nonexistent, confined to one gender, biparental, or communal. When parental care is offered, it may take such varied forms as oral or gill brooding, use of natural or constructed nests, internal gestation by a pregnant mother or by a pregnant father, or open-water guarding of fry.

Most genetic studies of mating behaviors in fish have been conducted on species displaying parental care of offspring. In the bony fishes (Osteichthyes), approximately 89 of the 422 taxonomic families (21%) contain at least some species in which adults provide direct postzygotic services, and in nearly 70% of those families, the primary or exclusive custodian is the male. Parental care by males alone is otherwise extremely rare in the vertebrates other than anuran amphibians. Thus, the evolutionary elaboration of paternal devotion makes fishes particularly favorable for testing traditional parental-investment and sexual-selection theories originally motivated by research on mammals and birds, where females typically are the primary caregivers.

Also intriguing for genetic analysis are alternative reproductive tactics (ARTs) *within* a species, or sometimes even within an individual during its lifetime. An example of the latter occurs in sequential hermaphroditic species in which an individual fish may switch its gender (and associated mating behavior) from female to male or vice versa. Both sex-changing and nonchanging fish are present in some populations.

Most fish species have separate sexes (i.e., are gonochoristic), and, within a gender, ARTs may be prevalent. In theory, a male fish may maximize the number of eggs he fertilizes by being quicker than rivals in "scramble competition," monopolizing mates or resources such as nests or territories, exploiting the resources of other males via reproductive parasitism, or cooperating or trading with resource holders via mutualism or reciprocity. Two or more such tactics are often observed in a population. For example, four types of males co-occur in the ocellated wrasse (*Symphodus ocellatus*): large "bourgeois" males that build nests and tend progeny; small males that parasitize (or cuckold) a bourgeois spawner by sneaking into a nest and "stealing" some of the fertilization events; medium-sized males that defend another male's nest from sneakers, but also court females and occasionally spawn; and extra-large males (pirates) that temporarily usurp the nest of another male.

In another example, male salmon spawn either as full-sized anadromous adults after returning from the sea, or as dwarf precocious parr that have remained in freshwater. In marker-based parentage analyses of Atlantic salmon (*Salmo salar*), parr have been shown to fertilize widely varying proportions (5–90%) of the total eggs in various populations. They also produce physiologically superior spermatozoa, a feature that partially compensates for their behavioral subordinance to dominant anadromous males.

The ARTs of anadromous salmon and resident parr appear tied to an individual's environmental exposure, but ARTs in some species might be genetically hardwired. Rearing fry under controlled conditions can help in evaluating developmental plasticity, as has been demonstrated with respect to alternative trophic morphs in several fish species. Regardless of their mechanistic basis, ARTs are common in fish, their occurrence facilitated by the prevalence in this group of external fertilization, a high incidence of paternal investment, and extensive intrasexual size variation attendant with indeterminate growth.

On the female side of the ledger, strategies of mate choice and parental investment can also vary. For example, in the peacock wrasse (*Symphodus tinca*), a female normally spawns with territorial males who care for her eggs on the nest, but she may also spawn off-territory with males who provide no parental care. Under some ecological circumstances, the cost of nest searching by a female may outweigh the lower survival of her untended offspring from these latter matings. Other ARTs known in female fish include additional variations on patterns of mate choice, parental care, resistance to coercion, and mating mode.

Although such ARTs are well described in the fish behavioral literature, their fitness consequences cannot be fully assessed from field observations alone. Genetic markers can shed new light on the realized success of ARTs in nature by disclosing actual biological parentage.

...

Three broad developments in the 1980s set the stage for refined genetic appraisals of fish reproductive activities. First, highly variable satellite DNA regions were found to be common features of eukaryotic genomes, and their utility in parentage assessment was quickly appreciated. Second, revolutionary insights came from analogous genetic studies in other taxonomic groups, notably insects and birds. For example, ornithologists supposed that most songbirds were genetically monogamous within a breeding season, but the new genetic data often excluded a nest-tending adult as the sire or dam of some nestlings. Such findings led to the realization that extrapair copulations and other clandestine reproductive activities, including conspecific nest parasitism, are routine phenomena in many avian species.

A third development was the realization that patterns of genetic parenthood are important for theoretical models of behavioral evolution. For example, surreptitious cuckoldry can yield a "genetic mating system" that departs from the apparent "social mating system" of a population, and this can impact the intensity of sexual selection and the evolution of secondary sexual traits. Furthermore, genetic analyses of birds, insects, and other taxa have suggested that realized parentage can reflect postcopulatory processes too, including sperm competition and female sperm choice.

Microsatellites are well-suited for genetic parentage analyses because the DNA-repeat units are small (2–5 bp each) such that alleles separated through suitable gels can be identified cleanly; the assays are applied locus-by-locus so the data can be interpreted in simple Mendelian terms; the polymerase chain reaction (PCR) is employed, so data can be recovered from even small amounts of tissue such as a single fish embryo; and allelic variation in most fish populations is extremely high.

A variety of statistical approaches for parentage analysis have been developed for particular biological settings. However, the basic logic of parentage analysis is generalizable. One common situation in fishes is when the male parent of a brood is known or suspected from genetic or behavioral evidence, and maternity is in question. For each offspring at each locus, the maternal allele can be deduced by subtraction (except when the sire and offspring are identically heterozygous). Then, any female whose genotype is inconsistent with these maternally deduced alleles at multiple loci is excluded as the dam.

An average exclusion probability refers to the mean probability of excluding an unrelated adult as a parent of a randomly chosen juvenile. In nearly all cases considered in this review, genetic markers were sufficiently variable that mean multilocus exclusion probabilities were well above 0.95. Such exclusionary power may earmark the true dam, but this also depends on the number of candidate females in the population, the thoroughness with which they have been sampled, and their genetic relationships.

Similar logic applies to paternity exclusions in nest-tending fish when particular offspring display alleles incompatible with those of their male custodian. Cuckoldry (stolen fertilizations by other males), nest piracy, and egg thievery are among the behavioral possibilities that may account for such cases of male foster parentage, and these can often be distinguished by considering details of the natural-history setting. Finally, when neither fish parent is available for genetic examination (as is normally true in species lacking parental care), the statistical exclusionary power and, hence, the capacity to draw biologically informative conclusions, usually is reduced considerably.

In most of the genetic appraisals of parentage and reproductive behaviors in fishes published to date, discrete cohorts of embryos within a nest (or inside a gestating parent) typically were genotyped in conjunction with a custodial adult and other individuals sampled nearby. By straightforward chains of reasoning, these multilocus genotypic data permit powerful deductions about the genetic parentage of particular juveniles, and such information accumulated across hundreds or thousands of fish from multiple nests can reveal the relative success of ARTs as well as the genetic mating system of a natural population.

What follows are brief synopses of various natural-history phenomena illuminated via genetic parentage analyses in fishes, in particular, species with parental care of offspring.

North American Sunfishes

In nearly all 30 species of North American sunfish (Centrarchidae), males guard eggs and embryos in shallow depression nests swept in the soft substrate of a lake or stream. Mart Gross and his colleagues have made one species—the bluegill, *Lepomis macrochirus*—a model system for the study of ARTs.

Cuckoldry by Males

Various routes to paternity are available to bluegills at the study sites in eastern Canada. "Parental" or bourgeois males, which mature at seven years of age, construct nests in colonies, attract females, spawn with them on the nest, and vigorously defend the nest and embryos against intruders. Precocious cuckolder males, by contrast, attempt to steal fertilizations from nest holders. These cuckolders are often 2- to 3-year-old "sneakers" that dart into a nest and release sperm as the bourgeois male spawns with a female, or older "satellites" that mimic females in color and behavior but release sperm as the primary couple spawns. Cuckolders leave the nest after spawning and show no parental care. They represent "alternative life histories" to that of bourgeois males.

What fraction of the reproductive output is attributable to bourgeois versus cuckolder males? Molecular markers provide the answer. In the largest genetic study of bluegills, involving 38 nests in one colony, the percentage of offspring per nest sired by the resident male ranged from 26 to 100 (mean, 79%). Cuckoldry by neighboring bourgeois males was rare, so about 20% of the embryos were the result of fertilization thievery by satellites or sneakers. The levels of cuckoldry per nest also were evaluated in conjunction with behavioral observations. The comparisons showed that as bourgeois males detect paternity lost to cuckolders (by assessing intrusion rates of sneakers and perhaps by using olfactory cues on offspring relatedness), they adaptively lower their level of parental care.

The overall rate of cuckoldry in this population closely matched the observed proportion (20%) of males at age 2 destined to become cuckolders. This raises the possibility that the genetic fitness of an individual may be similar for bourgeois males and cuckolders, a finding consistent with the notion that these two ARTs might be near an equilibrium frequency in an evolutionarily stable system perhaps maintained by frequency-dependent selection. This conclusion remains tentative, however, because snapshot appraisals of genetic parentage do not yield estimates of lifetime fitness. Furthermore, although not necessarily crucial to the question of evolutionary stability, it remains uncertain whether the ARTs in bluegill reflect a genetic polymorphism or a conditional ontogenetic switch regulated by the social or environmental experiences of a male during its development.

My colleagues and I have used microsatellites to assess paternity rates for bourgeois males in four other centrarchid species (*Lepomis auritus, L. punctatus, L. marginatus,* and *Micropterus salmoides*), and the results show that most or all embryos in a majority of nests were sired by the nest attendant. Mean cuckoldry rates were about 2%, or roughly an order-of-magnitude lower than in *L. macrochirus.* Two factors probably contribute to this difference. First, at the sites studied, the other centrarchid species were either solitary nesters or less colonial than the bluegills assayed, and, all else being equal, lower nesting densities probably reduce the opportunities for cuckoldry. Second, specialized cuckolder morphs are not known in *L. auritus, L. marginatus,* or *M. salmoides,* and although such morphs have been reported in *L. punctatus,* they were rare at the study site.

Nest Takeovers

Among the total of 142 centrarchid nests genetically surveyed, the custodial male in six instances (4.2%) had sired none of the young. One such instance involved a nest tended by a sterile F_1 hybrid between *L. macrochirus* and *L. gibbosus,* so the author concluded that this male had been 100% cuckolded. However, most of the other cases (all intraspecific) probably reflected nest-takeover events, and these nest piracies account for the majority of documented foster parentage. Perhaps nest takeovers are opportunistic responses to limited nest-site availability, or, perhaps the captured "nest holder" at the time of sampling was merely a temporary visitor (e.g., was there to cannibalize embryos).

Cuckoldry by Females

Largemouth bass (*Micropterus salmoides*) are unusual among fishes for tendencies toward biparental care of young and for staying with schooling fry for up to a month posthatching. Most of the 26 offspring cohorts genetically assayed in *M. salmoides* proved to be composed of full-sibs (consistent with the social monogamy suspected for this species), but four cases were documented in which the custodial female was the dam of most but not all of the juveniles that she and her mate tended. Most likely, a second female had laid some eggs in another's nest and then left her offspring in the care of their father and stepmother. "Cuckoldry" is normally meant to imply a reproductive behavior by which a breeding individual surreptitiously usurps parental services of another adult of the same gender, so this can qualify as a case of "female cuckoldry" analogous to male cuckoldry discussed above.

Multiple Mating by Bourgeois Males

In *M. salmoides,* the genetic parentage data show that successful spawning was usually by one female (and one male) per nest. However, in three other sunfish species similarly assayed—*L. auritus, L. marginatus,* and *L. punctatus*—the genetic data showed that multiple females typically had spawned with each bourgeois male. In other words, such nests contained mixtures of full-sib and paternal half-sib embryos. In the spotted sunfish, for example, the mean number of mothers per nest was at least 4.4, and statistical adjustments suggest that the true number occasionally must have been 10 or more.

Other Nest-Tending Species

In several other nest-tending fish species from both the freshwater and marine realms, parentage analyses by genetic markers have likewise been used to estimate numbers of dams per nest, rates of nonpaternity for bourgeois males, and cases of nest piracy. Additional reproductive phenomena have been uncovered as well, as described next.

Egg Thievery

Another behavioral route to nonpaternity for custodial males is egg stealing, a nest-raiding phenomenon occasionally observed, for example, in sticklebacks, Family Gasterosteidae. A bourgeois male uses kidney-secreted glue to construct a nest in vegetation, into which females lay eggs. Then the resident male (and, sometimes, sneakers) swim through the nest, releasing sperm. Occasionally, a bourgeois male is also seen transporting to his own nest a discrete cluster of eggs (clutch) that he has stolen from a neighbor.

Are these eggs viable, and had they been fertilized by the neighbor? Applying microsatellite markers to the problem, we have documented probable cases of egg piracy in about 17% of the nests in a population of fifteenspine sticklebacks (*Spinachia spinachia*). Each such instance was adduced when one of two or more clutches of viable embryos in a nest (rather than a few scattered progeny) had been sired by a male other than the bourgeois attendant. Using a slightly different DNA-fingerprinting approach, researchers in another lab similarly found that about 18% of nests in the threespine stickleback, *Gasterosteus aculeatus,* contained some stolen eggs.

Why would nesting males often pilfer fertilized clutches? Several hypotheses have been advanced for how natural selection on males might have promoted the evolution of egg-stealing tendencies: (a) the pump-priming effect—females in several species, including stickle-backs, are known to spawn preferentially in nests that already contain eggs; (b) the predator-dilution effect—extra eggs in a nest might ameliorate predation on the guardian's own embryos; (c) kin selection—if the larcenist and his victim are close genetic relatives but the thief has much higher prospects for successfully rearing offspring, both individuals might benefit from the theft in terms of inclusive fitness; or (d) the larder-stocking effect—males may steal eggs only to eat them later.

Egg Mimicry

In *Etheostoma* darters of the eastern United States, males of several species appear to have evolved bodily structures (typically on the tips of fins) that closely resemble eggs of these species and have been interpreted as "egg mimics." In a population of one of these species, *E. virgatum*, the egg mimics are displayed as pigment spots on the pectoral fins. In a genetic maternity analysis of fertilized eggs in the nests of 10 males, we found a significant correlation between the number of egg mimic spots on nesting males and their respective numbers of genetically deduced mates. Results are consistent with the hypothesis that egg mimicry by bourgeois males helps to attract gravid females to a nest.

Brood Parasitism by Helper Males

Cooperative breeding is fairly common in avian and mammalian species, but is known in only eight species of fish. In a nest-guarding cichlid fish from Lake Tanganyika, *Neolamprologus pulcher,* a pair of breeders often shares brood-care duties with individuals from previous clutches. In general, nest helpers might gain personal benefits such as food, protection, parental experience, or inheritance of a territory or mate, and/or they might gain in terms of inclusive fitness by rearing kin. Might they also profit in the immediate currency of personal fitness by siring some of the offspring within the brood? Yes.

Using multilocus DNA-fingerprinting assays, one team of researchers showed that about 10% of the progeny in seven assayed families of *N. pulcher* were fathered by helpers.

Reproductive Variance and the Opportunity for Sexual Selection

One common notion, supported by many studies of avian species, is that extrapair fertilizations enhance the opportunity for sexual selection by increasing the variance in male reproductive success. By stealing fertilizations from neighbors, some males become bigger winners (and others bigger losers) in the reproductive sweepstakes. Although this view may generally be true in socially monogamous species such as many birds, it may not hold in all situations. Namely, whenever the variance in reproductive success among males is larger in the absence of cuckoldry than in its presence, the opportunity for sexual selection may actually decrease with increasing levels of fertilization thievery.

Such may well be the case in the sand goby, *Pomatoschistus minutus*, a small European marine species in which males build and defend nests under mussel shells. Nest sites can be at a premium, and males often mate with multiple females. Thus, in total reproductive output, successful bourgeois males might be expected to greatly surpass other males in the population, especially those unable to secure nesting sites. However, as demonstrated genetically in our laboratory, fertilization thievery via sneaking is also extremely common in this species, occurring in about 50% of all nests. By interpreting these empirical findings in the context of models relating the intensity of sexual selection to variances in male reproductive success, we concluded that for this species, cuckoldry by sneaker males probably substantially reduces the opportunity for sexual selection.

Oral Brooders

Adults in many fish species protect their offspring by carrying eggs and hatchlings in the mouth or gill cavity. Oral incubation, in particular, is prevalent and has evolved many times independently in fishes of the Family Cichlidae. Microsatellite-based paternity analyses by various laboratories have documented multiple paternity of broods in several cichlid species, with up to six males fertilizing a single clutch.

Another otherwise cryptic phenomenon in mouth brooders, documented by microsatellite markers, is the shuffling of conspecific broods. In four of six orally brooded cohorts of fry examined in a Lake Malawi cichlid, *Protomelas spilopterus*, the proportions of juveniles not dammed by the female who held them ranged from 6% to 65%. Several possible explanations for the origin and significance of brood mixing remain highly speculative, but based on genetic and other evidence, this foster behavior in cichlids is remarkably common.

Female-Pregnant Species

In several fish groups, including the Poeciliidae (a large New World Family of live-bearers) and the Embiotocidae (the only Family of marine teleosts that is exclusively viviparous), a female is impregnated by one or more males and carries the resulting embryos internally, giving birth weeks or months later. Internal gestation guarantees that a pregnant female is the biological mother of her brood.

Multiple Mating by Females

In most nest-tending fishes, there is an inherent gender asymmetry in the genetic power to detect multiple mating. Each clutch typically is associated with a male guardian, so any multiple in situ mating by that male will be apparent in suitable molecular assays of his focal nest. However, multiple mating by a female can only be revealed if separate nests containing her progeny were included in the field collection, and this may seldom be the case when populations are large or sparsely sampled. Thus, if only for this bias, multiple mating by females has rarely been genetically verified in nest-tending fishes. However, this detection bias is reversed in female-pregnant fishes, where multiple mating (if present) by females is normally far easier to document genetically than is multiple mating by males.

In the sailfin molly, *Poecilia latipinna,* researchers revealed through allozyme-based paternity analyses that at least 52% of assayed broods were composed of embryos sired by two or more males, and that larger females were the more likely to have had multiple mates. In another example, at least 56% of pregnant mosquitofish (*Gambusia affinis*) carried broods of mixed paternity. However, marker variability itself can affect such estimates, as suggested by a later microsatellite analysis of mosquitofish in which we documented multiple paternity in nearly 100% of the surveyed broods.

Sperm Storage by Females

Ovarian tissues in poeciliids can store functional sperm for at least 1–2 weeks postcopulation, but storage of viable sperm by female surfperches (Embiotocidae) routinely occurs across several months. Using allozyme assays, researchers have shown that most broods in the shiner perch, *Cymatogaster aggregata,* are sired by multiple males, despite the fact that the matings preceded fertilization by 25 weeks or more (thus evidencing long-term sperm storage by females). This may be the longest known duration in fishes for potential sperm competition and postcopulatory female choice. However, even this pales in comparison to the multiyear utilization of female-stored sperm that has been genetically documented in some turtles.

Biological Benefits of Female Promiscuity

In female-pregnant fish, promiscuous mating tendencies by the males are evident in their vigorous sexual behavior and are easy to understand, but why would females also mate promiscuously? Multiple mating may expose a female to higher risks from sexually transmitted diseases, predation, copulation brutality, or other time or energy expenses associated with the mating process, and these potential costs might seem to outweigh any benefits in genetic fitness. However, a female fish might in principle gain any of several fitness advantages by mating with multiple males, including fertilization insurance against male sterility, access to more or better quality territories, success in "prospecting" better genes for her progeny, production of broods with more diverse and potentially adaptive genotypic arrays, and avoidance of inbreeding depression if some of her matings might be with close kin.

Female guppies, *Poecilia reticulata*, often solicit matings from multiple males, and many broods have multiple sires, especially in high-predation regimes. In microsatellite-based paternity analyses, researchers have discovered that females who had mated with multiple males had shorter gestation times and produced larger broods containing progeny with better-developed schooling behaviors and predator avoidance. These findings provide some of the first experimental evidence in fishes that promiscuity can be genetically rewarding for females as well as males.

Male-Pregnant Species

Two motivations have guided most genetic parentage analyses in fish: intellectual curiosity about a species' natural history and a desire to test broader mating system theories. Nowhere has the latter objective been more evident than in recent molecular appraisals of the Syngnathidae. A universal feature in the more than 200 living species of pipefishes and 30 species of seahorses is male pregnancy. One or more females lay eggs into a male's brood pouch or ventral surface, where they are fertilized by the assured sire and then housed as developing embryos until parturition weeks later. Such high paternal investment in offspring, and a freedom of parental responsibility for females, contrast diametrically with the situation in most mammals and many birds, making the syngnathids ideal subjects for testing, from a mirror-image perspective, traditional notions about gender roles in the context of mating system theories.

Mating Systems, Sexual Selection, and Sexual Dimorphism

In the behavioral literature on syngnathid fishes, "sex-role-reversal" is usually defined not as male pregnancy per se, but rather as any situation in which females compete more intensely for access to mates than do males. By this definition, some syngnathids are sex-role-reversed and some are not. In other words, in some but not all syngnathid species, females potentially produce more eggs during a breeding season than the available brood pouches of males can accommodate, such that males are the limiting resource in reproduction. This situation differs from that in most nest-tending teleosts, where rates of egg care by guardian males usually exceed rates of egg production by females.

The reason for defining sex-role-reversal in this fashion (whether stemming from male pregnancy, or from any other impacts on the relative reproductive rates of the sexes) is that the phenomenon then ties rather directly to broader theories on mating systems and sexual selection. Namely, because sex-role reversal produces a female-biased "operational sex ratio," it presumably is associated with higher intensities of sexual selection on females, a greater potential for the elaboration of secondary sexual traits in that gender, and mating systems tending toward polyandry. All of these predictions fall on the opposite end of a mating-system spectrum from the polygynous behaviors that characterize, for example, many mammal and bird species with traditional gender roles. In these other organisms, males often have the potentially larger variances in fitness, compete actively for females (the limiting resource in reproduction), experience more intense sexual selection, and often express a greater elaboration of sexually selected behavioral or morphological traits.

In some syngnathid species, sex-role reversal has been evaluated experimentally as the potential reproductive rates of males versus females. In other syngnathid species, the evidence for or against the phenomenon is indirect, involving, for example, the observed degree of dimorphism in secondary sexual characters. In syngnathids, when one gender is more brightly colored or otherwise sexually adorned, it is, indeed, normally the female. Given that syngnathid species appear to vary considerably along the sexual-selection continuum, conventional theory suggests that their mating systems may also vary accordingly.

In initial tests of this hypothesis, we have recently conducted microsatellite-based appraisals of the genetic mating system in each of five syngnathid species that display differing degrees of sexual dimorphism. The results proved to be in general agreement with broader mating system theory in at least two regards. First, the genetic mating systems fell along the monogamy–polyandry end of the mating system continuum, rather than in the monogamy–polygyny range of the spectrum as is normally true in most mammals and birds. Second, the strongly sexually dimorphic pipefish species (*Syngnathus scovelli* and *Nerophis ophidion*) proved to be genetically polyandrous, whereas a seahorse species (*Hippocampus subelongatus*) in which males and females show no elaboration of secondary sexual traits was genetically monogamous within a breeding episode. Furthermore, two pipefish species (*S. typhle* and *S. floridae*) that are intermediate in level of sexual dimorphism displayed a polygynandrous genetic mating system in which both females and males probably had multiple mating partners during the course of a male pregnancy.

These initial genetic findings for syngnathids conform to the general expectations of sexual-selection theory and mating-system evolution as applied to taxonomic groups containing role-reversed species. Caution is warranted, however, because many proximate ecological factors (as well as phylogenetic constraints) may also influence mating systems. In only a few fish species have molecular parentage analyses been applied to two or more populations, and pronounced geographic variation in the genetic mating system sometimes has been present, but at other times it has not.

Sex-Role Reversal and Bateman's Gradients

In the literature on animal-mating systems, the relative intensities of sexual selection on the two genders has been variously attributed to differences in parental investment, operational sex ratio, relative variances in reproductive success, and potential reproductive rates of the sexes, among others. Although such factors certainly impact the nature of sexual selection on males and females, A. J. Bateman argued more than 50 years ago that they all do so via one common denominator or first-order factor: the average relationship between the number of mates an individual obtains (its mating success) and the number of offspring that it produces (its reproductive success or genetic fitness).

Working on experimental populations of *Drosophila*, Bateman noticed that males' mean genetic fitness tended to increase rapidly with mating success (yielding a steep, linear selection gradient), whereas females gained little in offspring counts by mating with multiple males (a shallow or flat selection gradient). Bateman saw this disparity as the true cause of differential sexual selection; multiple mating afforded to males a higher fitness payoff than it did to females. More recently, "Bateman's gradients" have been touted by some authors as quantitative keys to comparing the intensities of sexual selection across species as well.

In sex-role-reversed taxa, steeper slopes in Bateman's gradients are predicted for females than for males (the reverse of the usual situation in *Drosophila*, and in many mammals, birds, and other taxonomic groups). Using microsatellite-based paternity analyses to assay the reproductive success of genetically marked individuals, we have critically tested this expectation using aquarium populations of a role-reversed pipefish species, *S. typhle*. Consistent with theory, the sexual selection gradients proved to be significantly steeper for females than for males. Results supported the Bateman-gradient approach for characterizing the strength and direction of sexual selection, and its underlying notion that the relationship of mating success to fitness is a cardinal feature in the process of sexual selection.

Extreme Polyandry and Sex-Role Reversal

One of our studies focused on the gulf pipefish *S. scovelli*, for which we uncovered perhaps the most compelling data yet available for any vertebrate species that sexual selection in nature can act more strongly on females than on males. One small population from a well-demarcated patch of seagrass was sampled exhaustively, thus enabling more complete evaluations than are normally possible on rates of genetic parentage by individuals of both genders. From genetic maternity and paternity analyses of the 21 broods, each pregnant male had mated with only one female, but on average a female had mated with 2.2 males. Furthermore, the standardized variance in female mating success (the variance in the number of embryos dammed by females, divided by the square of the mean—a gauge of the opportunity for sexual selection), proved to be at least seven times greater than the standardized variance in the mating success of males (including those not pregnant). This may represent the highest female-biased asymmetry of reproductive roles yet documented in nature for any vertebrate species, including several of the well-known shorebirds with sex-role reversal such as the phalaropes and jacanas.

Phylogenetic Character Mapping

A popular approach in recent years is to trace the evolutionary origin and modification of particular morphological or behavioral features through species' phylogenies estimated independently from molecular or other evidence. For example, a cladogram for the Syngnathidae, based on mtDNA sequences, was recently generated and used as a phylogenetic backdrop for interpreting the diversification of varied brood pouch morphologies within the Family.

This same phylogeny also provided a foundation for interpreting genetic paternity data in the context of the evolutionary rationale for brood pouch elaboration. *Nerophis ophidion* is unusual among syngnathid species in that adult males fertilize eggs externally and carry the resulting embryos on the outside of their bodies, rather than in an enclosed brood pouch. This arrangement opens a possibility for fertilization thievery by other males. Nonetheless, paternity analyses based on microsatellites showed that cuckoldry in this species is rare or nonexistent. The basal lineage leading to *N. ophidion* branched off early in syngnathid family tree. Thus, the genetic paternity data suggest that the evolutionary elaboration of enclosed brood pouches in other species of pipefishes and seahorses probably was not in response to strong selection pressures on pregnant males to circumvent cuckoldry, but rather as a means to enhance offspring care and protection.

Broadcast Spawners

Most fish species provide no parental care to their offspring. Typically, the dispersed fry from a spawning event do not remain associated with particular candidate sires or dams, so parentage assessments are far more challenging and problematic. Nonetheless, such focused genetic analyses have proved fruitful under some circumstances.

For example, we used allozyme markers to test the hypothesis that schooling juveniles of an open-water spawning coral reef fish, *Anthias squamipinnis,* had remained together throughout the pelagic dispersal stage and settled onto a reef as full-sib cohorts. The genetic data proved that juveniles in each school were not close relatives, but instead were a random draw from the local gene pool. In a similar study of a European minnow, *Phoxinus phoxinus,* other researchers likewise showed that discrete schools consisted of unrelated individuals. Conversely, microsatellite markers have revealed that discrete fish shoals in the tilapia, *Sarotherodon melanotheron,* often consist of closely related specimens.

In summary, fish parentage analyses based on microsatellites or other molecular markers have unveiled many facets of reproductive natural history and mating systems that would be difficult if not impossible to illuminate by other means.

14

An American Naturalist's Impressions
on Australian Biodiversity

Biodiversity in the modern age is under threat, ultimately from pressures associated with human overpopulation. In this chapter, contrasts are drawn between how two similar yet very different continents may be facing conservation challenges related to human numbers.

In 2001, a symposium entitled "Biodiversity Conservation in Freshwaters" was held in Canberra, Australia. The speakers addressed root causes for the imperiled state of freshwater biodiversity in Australia, and the resulting management challenges. Australian wetlands and watersheds, already scarce, have come under assault in recent decades from increased water usage by humans and the introduction of exotic species. Speakers representing management agencies and academic units generally were saddened and alarmed by the decline of Australian biodiversity, and by a perceived inadequacy of governmental and societal responses to the problem. At the close of the conference, organizers asked the handful of biologists visiting from overseas for our initial reactions to the Australian conservation experience, and how the challenges might compare with those on our own continents. Here are some of my impressions.

As first appreciated by Sclater and Wallace in the 1800s, Australia and North America are two of the planet's six primary biogeographic provinces. During the past 80 million years, Australia's geographic isolation has promoted evolutionary radiations of many characteristic biotic groups, including marsupial mammals, corvidlike birds, and eucalyptus and acacia plants. Humans colonized Australia about 56,000 years ago, and this may have contributed to the extinction of much of the continent's native megafauna (land animals weighing more than 45 kg) that began shortly thereafter. In North America, a similar mass extinction followed the much later arrival of humans there, about 13,000 years ago. Thus, people have long impacted indigenous biotas on both landmasses. Furthermore, after the arrival of Europeans on both continents in the past few centuries, the pace of environmental modification has accelerated tremendously and now threatens many kinds of creatures, both large and small.

The conference kicked off my second three-week-long visit to Australia, the first having been 10 years earlier. Although brief, these trips included wildlife excursions through the eastern and central portions of the continent—New South Wales, Tasmania, Queensland, South Australia, and the Northern Territory (including Arnhem Land and the interior near Alice Springs). On both occasions, I fell in love with Australia's rich native fauna and flora, the country's relatively unspoiled natural landscapes, and, most of all, the uncrowded feel of the continent. To a natural historian born and raised in the more congested United States, my first impression is that Australia's land and nature are about as untrammeled as those in North America may have been approximately 150 years ago—impacted and far from pristine, but not yet overwhelmed by human presence.

Of course, Australian biodiversity *is* under heavy siege from human activities. With regard to aquatic habitats, conservation issues highlighted at the conference were the urgent need to protect Australia's imperiled river drainages and aquifers, to properly identify and manage freshwater biodiversity on regional as well as local scales, and to control the introduction and spread of nonindigenous species such as carp fish, which often wreak havoc on native wildlife and ecological processes. Problems also abound in the terrestrial realm. For example, one response to the limited availability of agricultural land in Australia has been to forage cattle across much of the continent, and this has transformed the land. Feral and exotic species such as cats, rats, mice, goats, rabbits, and foxes also have had a huge impact by predating or competing with native wildlife. Some Australian scientists have been especially active in promoting the release of genetically modified (GM) organisms as a potential means to control exotic pests. For example, GM viruses have been engineered to serve as a vector for transgenes that induce rabbit sterility.

Despite the serious scientific and social challenges that conservation efforts in Australia clearly face, I couldn't help but conclude, during the Fenner Conference and in my travels, that the Australian natural environment and its native flora and fauna remain in far better shape than those in most of North America. In much of the continental United States, remnants of nature tend to be confined to small pockets in a human-dominated landscape, whereas in Australia, nature more often tends to be the basal matrix in which islands of human modification are embedded. I say this not to diminish any sense of urgency or commitment to Australian conservation initiatives. To the contrary, by comparing human population pressures on the two continents, I hope to remind Australians about how much natural richness they currently retain (and potentially stand to lose), the timeliness of the task, and the likely eventual outcome if Australia follows the historical paths of overpopulation and environmental disregard that too often have characterized North American society. Sadly, in the United States, in particular, great ecological opportunities were lost during the critical nineteenth and twentieth centuries as the country directed its capitalistic energies on shortsighted economic plunder, too often unduly heedless of the continent's environmental heritage as well as its future needs.

Australia and the continental United States are strikingly similar with respect to geographic area (7.7 and 7.8 million square kilometers, respectively), latitudinal arrangement (mostly 15–40° S vs. 25–45° N), and even geophysical layout. They are strikingly different, however, in terms of their current human population sizes.

In the vast western zone of Australia, Perth is the primary munici-
pality with slightly more than 1 million people. In contrast, several
far more populous urban areas are situated in the western United
States, including Los Angeles with about 9 million inhabitants. In the
low-latitude central zone of Australia, Darwin is the only metropoli-
tan center and contains only 70,000 inhabitants, whereas about 60
cities in the corresponding U.S. region exceed 50,000 residents.
Among these are Dallas (1.8 million people), Houston (1.6 million),
and San Antonio (0.9 million). In the eastern temperate zone, New
South Wales and Victoria are by far the most populous Australian
states, yet their total of 10.7 million citizens is little more than the
number of people who inhabit New York City proper, which is just
one of several dense epicenters in a quasi-continuous megalopolis extending from Boston to Washington, DC, and beyond. In the high-lat-
itude central zone, Australia has little more than Adelaide with 1.0
million people, in contrast to that same region in the United States
that houses Chicago (nearly 3 million) and many other metropolises.

In short, by Australian standards, the United States is remarkably
crowded. About 300 million individuals are crammed within U.S.
boundaries, or about 15 times more people than in all of Australia.
The states of Texas and New York alone, with approximately 20 million residents each, both have about as many citizens as live on the
entire Australian continent, and California by itself houses nearly
twice that many individuals! Overall, population densities on the two
continents average about 37 versus 2.5 people per square kilometer.

In terms of human impact on the natural environment, this disparity in population densities has dramatic consequences. Many examples could be cited. To pick just one, on both continents a splendid
eastern peninsula juts far into tropical or subtropical oceans. Australia's Cape York is a sparsely inhabited wilderness, whereas America's Florida Peninsula is rimmed with condominiums, crisscrossed by
congested highways, and packed with more than 16 million people
vying for space and water with what remains of the native wildlife.

Of course, native landscapes and biodiversity are impacted not just by human numbers per se, but especially by per capita resource consumption and patterns of land use. Despite the environmental pressures of an exploding U.S. population during the past 150 years, greater foresight surely would have short-circuited many of the ecological crises that North America now faces. Today, we can only wish that our forebears had seen a need to set aside more extensive nature reserves; to cherish rather than pillage the land and its wildlife; to stem the introduction and spread of exotic species; to conserve nonrenewable resources ranging from topsoils to natural landscapes to underground aquifers; to manage renewable resources such as forests, grasslands, and wetlands; to place wiser restraints on commercial fishing practices; to promote clean air and water; and to think and plan in advance on patterns of urban development. Timely action could have ameliorated many of the detrimental impacts of humanity on the natural environment, on ecological processes and services, and ultimately on our own well-being.

From my travels, I have the distinct (but perhaps naïve) impression that Australian sensibilities, both within and outside of government, are sometimes far ahead of the United States on such matters, notably in the realm of urban development. Perhaps Australia is benefiting from perspectives promoted by the conservation movement of the twentieth century, but in any event it has the distinct advantage of still having a relatively low population density. Even Sydney, Australia's most populous city, has, within an hour's travel, no less than seven biodiversity-rich federal parks—including Royal, the world's oldest national park (formally recognized in 1879, four years before Yellowstone was designated a National Park in the United States). By contrast, to reach the nearest substantial parks under federal protection, a cement-weary New Yorker must travel about 500 kilometers, either to the Shenandoah National Park in northern Virginia or Acadia in Maine. Another striking contrast involves our nations' capitols—the carefully planned city of Canberra, richly adorned with parklands and open spaces, and surrounded by thinly populated countryside, exudes a nature-friendly ambience dramatically unlike the crowded urban aura of Washington, DC.

Perhaps it is merely historical circumstance and good fortune that thus far has preserved much of Australia's biotic and scenic richness. An arid interior and a paucity of major rivers has meant not only that Australia's aquatic biodiversity is necessarily low (in comparison with North America), but also that human occupancy of the region has been constrained. Ironically, from the standpoint of native ecology and natural environments, the relative scarcity of freshwater habitat in Australia (a situation often lamented at the conference) may prove ultimately to be the continent's ecological salvation.

Australia today, with its "evolutionary heritage" still reasonably intact, is approximately where North America found itself in the year 1850, when the U.S. human population stood at 23.1 million. Had wise environmental practices been promoted in North America beginning then, imagine the richer floras and faunas that still might be available for us to enjoy today. The good news for Australia is that it's not yet too late to preserve most of what evolution there has produced.

15

The Best and the Worst of Times
for Evolutionary Biology

This is a peculiar time for the field of evolutionary biology. On the one hand, powerful new molecular tools have opened the entire biological world for genetic analysis, on exciting fronts ranging from detailed laboratory analyses of the genomic workings to numerous aspects of natural history and nature's operations in the wild. On the other hand, the subject matter of evolutionary studies—biodiversity itself—is under unprecedented human assault, worldwide. In this chapter, Avise addresses the mixed emotions that this conundrum evokes. This essay was the outgrowth of a lecture originally delivered in Germany to a specially convened audience of business leaders, artists, and other accomplished citizens who had made lifelong contributions to European society. One of topics touched on in this chapter is the field of genetic engineering. Readers wishing to learn more should consult the author's book: *The Hope, Hype, and Reality of Genetic Engineering: Case Studies from Agriculture, Industry, Medicine, and the Environment* (2004). Another topic raised in the chapter is the utility of phylogenetic reconstruction for deciphering the evolutionary histories of various organismal traits. For much more on that subject, see the author's book: *Evolutionary Pathways in Nature: A Phylogenetic Approach* (2006).

It was the best of times, it was the worst of times, it was the age of wisdom, it was the age of foolishness, it was the epoch of belief, it was the epoch of incredulity, it was the season of Light, it was the season of Darkness, it was the spring of hope, it was the winter of despair.

Charles Dickens (opening paragraph from *A Tale of Two Cities*)

These evocative sentiments from Charles Dickens's classic (1859) appeared in the same year as Charles Darwin's *On the Origin of Species*.

They also encapsulate the feelings of many natural historians about the state of evolutionary biology today. The recent elucidation of the sequence of all three-billion-plus nucleotide pairs in the human genome is an example of how this is the best of times for evolutionary biology. This achievement, which will stand forever as a milestone in the history of science, is a crucial step toward someday deciphering the metabolic and physiological functions of legions of proteins and RNA molecules encoded by the approximately 30,000 human genes. If even a modest fraction of the scientific and media hoopla surrounding the genome project proves justified, not only will medical breakthroughs accrue rapidly, but so too will conceptual revelations about how evolution works.

Large-scale genetic profiling (such as by microarray techniques and comparative genomic sequencing) will help to identify and characterize the genes that influence particular organismal traits at the levels of metabolism, physiology, and morphology. From such approaches the long-sought Holy Grail of evolutionary biology—a fuller understanding of the causal links from genotype to phenotype—will gradually be achieved. In the first century following Darwin and Mendel, the basic driving forces of evolution were elucidated through observations on natural history, and the fundamental principles of heredity were uncovered by monitoring patterns of genetic transmission of traits in a few species that could be bred under controlled conditions. In this, the second post-Darwin century, scientists finally have direct and universal access to the genetic mechanisms of evolutionary change, potentially in any species we wish to investigate.

In a 1968 science fiction film *2001: A Space Odyssey*, by Stanley Kubrick and Arthur C. Clarke, several astronauts and a supercomputer named HAL were sent on a grand mission to explore the solar system. In real life in the 1960s, some of the earliest molecular genetic approaches were introduced to evolutionary studies, and in 2001 the first complete sequence of the human genome was analyzed (with extensive computer assistance). These achievements ushered in an exploratory era of science nonfiction (it could be named *2001: An*

Inner-Space Odyssey) that is turning out to be as intellectually fascinating as the outer-space odyssey envisioned by Kubrick and Clarke.

Another source of promise (but also trepidation) began in the early 1970s when researchers constructed the first recombinant DNA molecules in a Petri dish. Building on this technological breakthrough, molecular geneticists have gained unprecedented powers to reshape life. Now researchers routinely identify genes for a variety of biological functions, modify and reassemble these genes in test tubes, insert the recombinant DNA molecules into living cells, and thereby swap genetic material freely among any living species. Hundreds of plant, animal, and microbial species have been engineered to carry designer genes from foreign sources, and many of these transgenic organisms already have played or soon will play huge roles in medicine, pharmacology, environmental bioremediation (e.g., cleaning up toxic wastes), animal husbandry, and agriculture.

Some prognosticators believe that the application of recombinant DNA methods to gene therapy and gene replacement (the repair or replacement of defective genes in the body) soon may lead to a revolution in the history of medicine comparable to the introductions of sanitation, anesthesia, and antibiotics and vaccines. If the new recombinant gene technologies live up to their early billing, the day might soon come when gene therapy can alleviate sickle cell anemia, heart disease, cancer, or various other human genetic disorders. Just as we may marvel at our forbears' fortitude in the dark ages before the advent of our modern medicine, our grandchildren may look back with marvel at our own fortitude in the era preceding the wide availability of gene therapies. Nonetheless, the technical hurdles remain daunting. Although several hundred experimental gene-replacement trials have been conducted within the past decade (involving a total of several thousand human subjects), few if any definitive medical success stories exist to date, and the entire discipline is under intense scrutiny by advocates and critics alike. That said, there are good reasons to believe that gene technologies will revolutionize medicine sooner or later.

Even more daring is the proposal that genetic engineering soon might be extended to cells in the human germ line. In contrast to somatic gene therapy, which directly affects only the individuals receiving the procedure, the intent of germ-line engineering is to alter the human gene pool in subsequent generations as well. Obvious candidates for germ-line engineering are alleles that produce terrible genetic disorders such as cystic fibrosis or Huntington's disease. Who will object if molecular means can be found to reduce human suffering by correcting such conditions? But, also, who will favor efforts to engineer into one's children germ-line genes for cosmetic features such as height or athletic ability, or perhaps for higher IQ?

In the first half of the twentieth century, several eugenics movements around the world exalted the notion that *Homo sapiens* could be bettered by selective breeding. Such efforts came to a nadir in Nazi Germany, where racial extermination was the purported means to improve humanity's gene pool. Purposeful germ-line engineering must therefore be preceded by extensive ethical discussion by a broad cross section of society. Such initiatives will also have to distance themselves from ill-conceived eugenics movements of the past.

Recombinant DNA technologies are a double-edged sword in other arenas as well. One dangerous possibility is the production of biological weapons. In principle, it would be quite easy for someone with nefarious motives to mix and match genes from different species and thereby engineer deadly microbes invulnerable to conventional drugs. Even well-meaning scientists might create ghastly strains accidentally. This sobering prospect recently became more plausible after a poliovirus was (deliberately) synthesized chemically ex nihilo by using genetic information from a publicly available database. Another frightening viral strain was engineered to contain a mixture of genes from the dengue fever and hepatitis viruses. In one more instance of alarming use of recombinant DNA methods, pathenogenicity was inadvertently enhanced experimentally in a mouse analogue of the human smallpox virus. Any release or escape of such deadly organisms could have grave consequences.

There are other alarming reasons to fear that the twenty-first century could be the worst of times for biology. Ecologists and natural historians are painfully aware that the subject matter of their devotion—biodiversity—is under assault worldwide as the continents fill with people. The collective weight of human activities is leading to the disappearance of wilderness. Atmosphere and oceans are being polluted, marine fisheries are collapsing worldwide, and wetlands and freshwater aquifers have shrunk dramatically. In short, Earth's renewable and nonrenewable resources are being tragically squandered. In the Amazon Basin, for example, which is famous for its rich biota, slash-and-burn fires are so numerous that their light is visible to astronauts in the space shuttle. Some of these astronauts have felt moved to speak in a deeply spiritual tenor about the beauty of the "blue planet" and to bemoan how we are despoiling this special, fragile place.

Experts agree that we currently find ourselves in the midst of one of the largest mass extinction sagas in the history of life. Species are being lost at rates at least 100-fold higher than they were before the coming of humanity, with total losses by century's end projected to be somewhere between 10% and 50% of the Earth's now-living biota. This biological holocaust has been unrivaled since the time, 65 million years ago, an asteroid struck the planet and precipitated a global winter. Biologists affected by biophelia—a deep emotional attachment to nature—grieve that biodiversity is now entering another winter of despair.

Against this backdrop of conflicting emotions about the state of modern biology, I want to describe four broad fronts where evolutionary science and its sister discipline of genetics face near-term societal, as well as scientific, challenges and opportunities.

During the Industrial Revolution that began two and a half centuries ago, advances in technology, in concert with wasteful consumptive practices, enabled humans to dominate the planet (and foul its life-support systems). Some might argue that this is sufficient ground on which to reject any new biotechnologies in the offing. Others, however, might agree with E.O. Wilson's recent assessment of the current biodiversity crisis when he wrote: "Science and technology led us into this bottleneck. Now science and technology must help us find our way through and out."

Whatever one's sentiments about recombinant DNA methods, this genetic genie is already well out of the bottle. In the early 1970s, fewer than 20 years after the discovery of the structure of deoxyribonucleic acid (DNA), scientists first transferred foreign genes into *Escherichia coli* and coaxed these genetically modified (GM) bacteria into producing valuable medical compounds, such as human insulin. These scientists thereby established a path to commercial genetic engineering. A decade later, researchers created the first GM crop (transgenic tobacco), a feat that led to an ongoing revolution in plant genetic engineering. Today, patents are issued routinely for GM products and technologies in a wide variety of lucrative pharmaceutical and agricultural enterprises around the globe.

About 10,000 years ago, our forebears invented agriculture. By sowing the seeds of edible wild plants, harvesting the resulting foods or fibers, and retaining seeds from the best specimens for subsequent planting, they began to transform native plant varieties into the bountiful domestic fruits and vegetables of today. Such artificial selection over the centuries required no sophisticated gene technology, no detailed understanding of hereditary mechanisms, no cognizance of evolutionary processes—just a keen eye for desirable plants, strong arms to tend the crops, and patience. Today's agricultural engineers still need a keen eye for their subject, but patience no longer is necessary. Through recombinant DNA techniques, the genes of crop plants (and animals) can be manipulated directly and nearly overnight. Some people see an ethical imperative for such efforts, pointing out the burgeoning number of human mouths to feed. Others are outraged by such genetic manipulations and caution scientists not to interfere with their food.

The first GM crops were approved for commercial planting in the United States in the early 1990s, and within a decade roughly 50% of the corn, soybean, and cotton planted across the United States was genetically modified for one trait or another. At least 70% of the processed foods on American grocery shelves now contain ingredients derived from transgenic sources. Most GM crops in cultivation today were intended to improve food quality or to display resistance to disease microbes, insect pests, or herbicides. In general, consumers in the United States have accepted this transition, but public outcries against transgenic "frankenfoods" and agricultural "farmageddons" have been loud in much of Europe, causing those governments to block the spread of GM technologies. The different public reactions serve notice that societal attitudes as well as the science can play a huge role in the success or failure of commercial GM enterprises.

The available scientific evidence often leaves ample room for polarized opinions on GM crops, especially with regard to environmental issues. Among the potential blessings of GM crops are increased yields per acre, nutritional and health benefits to people and domestic animals, and ecological payoffs such as a rapid phaseout of dangerous chemical pesticides (for example, when GM crops genetically engineered for pest resistance are planted widely). Proponents of agricultural engineering argue that transgenic crops soon will usher in a "Gene Revolution" that will do even more than did the Green Revolution that began in the 1950s, when new varieties of high-yield crops were generated by more traditional plant-breeding methods.

There are, however, important scientific concerns. Transgenes in some GM plants could pose human health risks, for example, by inducing allergies. Another risk is that transgenes might escape to nontarget plants and precipitate ecological or agricultural disasters. Consider, for example, if transgenes for herbicide tolerance or insect resistance were to transfer from engineered crops into related weed

species. Social implications must also not be forgotten. Notably, the
widespread deployment of GM crops most likely will be attended by a
further shift toward monocultural farming practices, a diminution in
the number of indigenous crop varieties, and greater reliance by
farmers on large agribusinesses. Thus, agricultural shifts prompted
by the gene revolution will entail profound economic, social, and ecological consequences, and these may not always be for the good.

The most widely planted GM crops to date have been engineered to
carry *Bt*-toxin genes, named for the bacterium from which they derive—*Bacillus thuringiensis*. These microbial genes confer resistance
against particular insect pests. Ideally, these GM crops should alleviate much of the need for synthetic chemical insecticides, such as DDT,
that were a hallmark of the Green Revolution but also poisoned the
land and wildlife. On the other hand, potential biological downsides
to *Bt*-engineered crops include the possibility that transgenes might
leak into nontarget populations via pollen or seed flow or into other
species via introgressive hybridization, that the toxic *Bt*-proteins
could harm beneficial or other nontarget insects that feed on the
transgenic crops, or that populations of some targeted insect pests
might evolve genetic resistance to the transgenic *Bt* toxins and thereby render them ineffective.

It seems clear that the ecological and evolutionary-genetic sciences
can constructively inform the ongoing efforts of the high-tech agricultural industry. For example, appropriate experimental and theoretical research could answer questions such as these: How far and
where do the pollen and seeds of particular transgenic crops move?
With what wild species might GM plants hybridize? Which nontarget
insect species are affected by the plant-expressed *Bt* toxins, how severely, and with what ecological consequences? What are the molecular and evolutionary-genetic routes to *Bt*-toxin resistance in pest populations, and how likely is such genetic resistance to evolve under the
novel selection pressures stemming from widely planted GM crops?
How and where might crop plants best be engineered to prevent the

evolution of pest resistance? How might GM crops best be deployed in space to mitigate potential ecological dangers?

Seldom are such scientific issues seriously addressed by the industries that stand to profit from the genetic engineering projects or by government agencies mandated to oversee and license the commercial operations. As a net result of this lack of input from the ecological and evolutionary sciences, societies unnecessarily court too many biological disasters.

Another prime example comes from medicine, where a frightening development in recent years is the widespread evolution of microbial resistance to powerful antibiotics such as penicillin. For decades, these compounds were disseminated all too readily by a medical profession that failed to foresee predictable evolutionary responses by the microbes. Antibiotic supplements in commercial animal feeds have been another source of selection favoring the evolution of microbial drug resistance. Now, a desperate and costly scramble is under way to identify new generations of antibiotic drugs that can offer people renewed protection against the resistant "superbugs." Even a rudimentary understanding of evolutionary genetic principles by the medical and agricultural industries might well have prompted wiser antibiotic practices and thereby avoided this crisis.

Quite apart from biotechnological applications per se, the molecular revolution in biology is also yielding unprecedented conceptual insights into basic evolutionary processes. A good example comes from transposable or mobile elements. In the early 1950s, Barbara McClintock discovered these "jumping genes" as they moved about the genome of corn plants, hopping routinely from one chromosomal site to another, often replicatively. The significance of these observations went mostly unappreciated at the time, but jumping genes later were found to be important and nearly ubiquitous features of eukaryotic cells. In 1983, Dr. McClintock was awarded a Nobel Prize for her work.

Various classes of jumping genes and their less-frisky evolutionary descendants and relatives have proved to be astonishingly abundant in most plant and animal species. They often make up more than 50% of the genome. In humans, for example, each cell contains more than 500,000 copies of one class of 300-base-pair sequences (known as Alu), and about 100,000 copies of a longer family of sequences that accounts for about 5% of our total DNA. Such nucleotide sequences generally have no known function apart from their own self-perpetuation. Most stem from "master copy" sequences that over time have given rise to vast numbers of derivative sequences. By replicatively dispersing themselves across the genome, transposable elements enhance their own prospects for transmission to the next generation.

Jumping genes are thus prototypical selfish genetic elements. They also can be described as miniature intracellular parasites. Through their tendency to induce mutations, and also, perhaps, from the sheer metabolic burden of their vast numbers, jumping genes commonly damage their hosts. The analogy to parasites is apt in another regard: Phylogenetic discoveries based on DNA sequence analyses indicate that some jumping genes—the retrotransposable elements—have close evolutionary ties to the family of infectious viruses that includes the causal agent of AIDS in humans.

Like any association between host and parasite, however, selection-mediated evolution sometimes works out symbiotic relationships between the participants. Growing evidence indicates that at least some former jumping genes have even been recruited over evolutionary time into activities beneficial to their host. These include the sponsorship of recombinational variation of immune-response genes, the formation of centromeric regions that help direct chromosome movements during cell divisions, the repair of chromosomal ends (telomeres) whose decay otherwise is associated with the aging process, and the promotion of gene duplications and other genetic alterations that in general provide important (albeit fortuitous) fodder for evolutionary innovation.

Jumping genes are merely one among several "nontraditional" types of genetic elements whose presence could scarcely have been imagined in the premolecular era. Also inhabiting the human genome are vast armies of noncoding DNA sequences known as introns that stand like sentinels between the coding regions of protein-specifying genes; battalions of repetitive DNA sequences, each composed of nearly identical DNA sequences aligned in closed rank; active promotors and regulator sequences that act like field sergeants, ordering about the squadrons of proteins and nucleic acid molecules that do the grunt work of cellular metabolism; and legions of pseudogenes, former genes that are no longer functional but clutter the genome like corpses on a battlefield.

Thus, the traditional genetic image of a genome densely packed with DNA that benefits the cell has turned out, upon close molecular inspection, to be a mirage. The protein-coding genes that prescribe much of our genetic health are scattered about the genome like tiny desert oases embedded in long linear stretches of what appears at first sight to be a noncoding genomic wasteland. The noncoding regions make up the vast majority of our total DNA. Protein-coding genes have been the traditional focus of medical research on inborn errors of human metabolism, but they constitute just a tiny fraction (about 2%) of our species' genetic heritage.

Not that the rest of the human genome is therefore mere junk. What was formerly termed "junk DNA" is actually a treasure chest of information about evolutionary process. An analogy to garbage is appropriate. In recent decades, anthropologists have come to view ancient garbage dumps near human settlements as wonderful sources of historical information. Likewise, biologists are beginning to appreciate that by rummaging through the piles of junk DNA in our cells, they can unearth information about genes' evolutionary lifestyles. To fully catalogue such historical genomic information is a major research challenge for the coming decades.

An emerging evolutionary view is that the genome is in many ways like an extended intracellular society of interacting genetic elements. Within each such microecosystem are multitudinous quasi-independent DNA sequences with elaborate divisions of labor and functional collaborations. Such sequences can also engage in evolutionary feuds stemming from hereditary conflicts of interest. Such intergenic conflicts are nearly inevitable in any species that engages in sexual reproduction, because unlinked genes in such species are partially autonomous (as a result of the vagaries of Mendelian transmission). Genes are not all inherited together but instead are segregated and re-sorted in sexual reproduction. Thus, genes tend to evolve replication tactics that enhance their individual prospects for survival and transmission.

The net result is that different pieces of DNA within an extended lineage continually play coevolutionary games. Their strategies often bear striking analogy to those observed among people partially bound in social arrangements. These can include collaborative efforts but also individual opportunism, group alliances but also cheating, and societal strictures against any unduly egoistic tendencies of the individual. Such societal metaphors for gene-by-gene interactions in evolutionary as well as in contemporary time are less than perfect, but they do evoke a more realistic and powerful image of molecular affairs than does the image of genes as relatively inert beads strung along the chromosomes.

One more example further illustrates the beauty and complexity of intragenomic interactions. Genes in the living cells of all advanced organisms ultimately trace back to several (and perhaps many) endosymbiotic marriages between unrelated microbes early in the history of life. The most famous of these nuptial occasions, which occurred more than a billion years ago, was the formation of an intimate cellular union between a purple bacterium and another microbe that bore the precursors of many of the genes now housed in each cell's nucleus. Following this intercellular wedding, some of the purple bacterium's genes gave rise to the genome of mitochondria. Most, however, were incorporated into the evolving nucleus of a primordial eukaryotic cell. These genes continue to collaborate today with DNA stemming from other ancient microbes that likewise participated in early endosymbiotic amalgamations.

These are well-documented evolutionary happenings. Other molecular events that are legacies of the original endosymbioses still occur within the cells of individuals. For example, a multitude of proteins encoded by nuclear genes continually migrate to the mitochondria, where they engage in exquisite molecular ballets with the protein products of mitochondrial genes to mediate production of chemical energy.

Ancient microbial matrimonies left other legacies. In any zygote or fertilized oocyte, most of the cytoplasm comes from the egg rather than the sperm, so mitochondrial DNA (mtDNA) is transmitted to offspring almost exclusively via females. This makes mtDNA a valuable genealogical marker for deciphering the matrilineal component of any animal pedigree, much as surnames provide markers of patrilines in many human societies. Data from mtDNA show that all modern human matrilines trace to expansions of our species that occurred when our forebears left Africa in fairly recent evolutionary times.

Being maternally inherited, the modern mitochondrial genome also retains a quasi autonomy that can bring it into evolutionary conflicts of interest with biparentally inherited nuclear genes. For example, from the selfish perspective of a cytoplasmic gene, it matters little if males are sterile or debilitated, because males are not a viable avenue for cytoplasmic transmission. From this evolutionary vantage, it is no coincidence that a disproportionate fraction of genes contributing to male sterility in many plant and animal species are housed in the cytoplasm. Because cytoplasmic genes are transmitted maternally, they behave as if they are rather indifferent to male well-being. They may even jeopardize the longer-term evolutionary health of a species by biasing families toward producing daughters rather than sons, sometimes dramatically.

We are coming to realize that Darwinian processes operate not only at the traditionally understood levels of the organism and kinship group but also on DNA sequences engaged in the evolutionary struggle for existence. These molecular-level Darwinian processes are intimately tied to sexual reproduction, because under Mendelian rules of heredity, unlinked genes have noncoincident transmission routes and, thus, quasi-independent evolutionary fates. Different pieces of DNA tend to evolve individualized fitness strategies as they collaborate but also jostle for successful passage through an extended organismal pedigree.

Genes may be considered to resemble miniature intracellular deities in their dominion over human affairs, yet they provide mechanistic as opposed to otherworldly explanations. Genes are physical rather than metaphysical entities, natural rather than supernatural, real rather than ethereal. They give every indication of having been fashioned not by the loving hands of a conscious engineer, but by an amoral (not immoral) evolutionary process—natural selection—that shapes life at several hierarchical levels, including that of the DNA sequences themselves. Like other evolutionary genetic forces such as mutation and recombination, natural selection has no consciousness, no code of conduct, no reflective concern about the consequences of its actions. Selection is a powerfully creative and directive force in biology, but it is as uncaring as gravity or lightning about organismal well-being.

The often-surprising consequences of natural selection are also being explored now in many other contexts, including aging and death. Why, from an evolutionary perspective, should genes ever dictate senescence and mortality for the individual?

Theoretical population biologists have shown that senescence and death are virtually inevitable evolutionary repercussions of organismal reproduction. Natural selection tends to act more forcibly on genes transmitted through young rather than old reproducers. As a long-term evolutionary consequence, older age classes in any population tend to become developmental repositories for genes with age-delayed deleterious somatic effects. In part, such genes accumulate simply because of weak selection against their loss. A related realization is that genes for aging are favored by natural selection whenever their beneficial effects at early stages of life outweigh deleterious effects later on. For example, any genes that predispose for bone calcification in adolescents might improve an individual's genetic fitness by strengthening limbs. Under the action of natural selection over the generations, these calcification genes would increase in frequency, even if they also happened to harden artery walls and thereby promote heart disease later in life. The net effect of such age-related natural selection is a tendency for the evolution of a marked acceleration of death probabilities with advancing age.

In short, aging and death do not violate some rule of evolution by natural selection. Rather, they exist because natural selection fails to prevent the evolutionary accumulation of disabling genes in the elderly. These are just a few examples of how the field of evolutionary genetics can yield objective insights into human conditions that in all prior ages fell under the purview of mythology, theology, and religion.

More than a quarter-century ago, Carl Woese and George Fox used DNA sequences from a small ribosomal RNA gene to infer that life on Earth is divided into three primary historical kingdoms or domains: the Archaebacteria (a category of bacteria that was one of the early forms of life), the Eubacteria (other bacteria), and the Eukaryota (animals, plants, fungi, and protists). Their glimpse at deep branches in the tree of life demonstrated the astonishing power of molecular data for reconstructing "phylogeny" (a term for the evolutionary history of life). Since then, the volume of recovered genetic information has grown exponentially, and systematists now routinely employ nucleotide sequences to estimate the relatedness of species at any degree of evolutionary separation. For example, a recent meta-analysis of DNA sequences from more than 5,000 genes was used to infer the approximate evolutionary ages of the most recent common ancestors for more than 300 species representing the major groups of placental mammals.

Within the next decade or two, scientists almost certainly will complete a near-exact reconstruction of the tree of life, including inferred phylogenies for all major taxonomic groups recognized among the 1.7 million described living species. These genealogical histories will have been derived mostly from genes transmitted vertically from parent to offspring, but instances of horizontal genetic transfer between organismal lineages, as mediated by retroviral movements or other means, will also be documented.

Some classes of DNA evolve so rapidly that they illuminate historical relationships even among individuals within a species. The applications range from paternity and maternity assignments to analysis of population separations often dating to the Ice Ages. Apart from the historical reconstructions per se, the molecular genetic appraisals also will reveal a wealth of behavioral and natural history information. Thus, this century will also see a further flowering of molecule-based natural history.

This assembly of the tree of life will stand as another of the grand achievements in the history of biology, at least comparable in importance to the human genome project. It will provide the historical backdrop necessary for almost all studies in comparative biology, from the basic to the applied. For example, details in the tree of life will enable researchers to chart the phylogenetic origins and evolutionary transitions of any anatomical or physiological feature. More practically, the tree will enable scientists to describe how biodiversity has changed over time, which will aid conservation efforts, and it will assist bioprospectors in searches for pharmaceutical or other valuable compounds from nature. In short, the tree of life will serve as a comprehensive roadmap for nearly all exploratory research in biology.

A 1998 report from the National Research Council concluded that the United States is conferring too many graduate degrees in biology for current societal demands. The report likened the fates of many recent biology graduates to planes circling an airport: Postdoctoral students

typically enter a long holding pattern before gaining clearance to land
a job in their chosen profession, and having landed, still face daunting
hurdles before they secure research funding. This description speaks
very poorly of our nation's priorities. We live in an era when there is a
need for *more* biologists and earth scientists to help inform decisions
on the complex challenges that are of utmost concern in agriculture,
medicine, and the environment. Well-trained biologists are essential,
as are science-literate religious and political leaders and a scientifically informed public. A compelling challenge for government is to
structure legislation and economic incentives in ways that will promote the biological sciences.

In a popular article published in a 1973 issue of the *American Biology Teacher*, the evolutionary geneticist Theodosius Dobzhansky
penned a famous phrase. He wrote, "Nothing in biology makes sense
except in the light of evolution." Dobzhansky was referring not
merely to the genealogical history of life, he was also alluding to how
the evolutionary sciences explain biological phenomena by using dis-
passionate reasoning and objective evidence that are thus expressly
divorced from faith in metaphysical causation. It is a great irony that
in this age of genetics and biotechnology, most people are grossly
ignorant about evolution and genetics or openly hostile to their implications. For example, only about 10% of Americans believe that
evolution occurs as an entirely natural phenomenon, and about one-
third of high school biology teachers reject the concept of evolution
altogether.

Two recent examples from the United States reveal how precarious
the enlightenment's hold sometimes appears. In Ohio, fundamental-
ist Christian groups have been hard at work lobbying legislators and
school boards to water down or even squelch the teaching of evolution in public schools. In Georgia, the Cobb County school board
recently directed that biology textbooks include a sticker that in
essence disclaims evolution. Similar attacks on evolutionary biology
are initiated almost every year. These creationist salvos are much like
those launched by William Jennings Bryan during the infamous 1925

"monkey trial" in Dayton, Tennessee. There, defense attorney Clarence Darrow staunchly, but unsuccessfully, defended the right of John Scopes to teach evolution in public schools, a right that was denied in a state law sponsored by Protestant fundamentalists.

Some ardent creationists would have us believe that the Earth is only 10,000 years old, that all species were forged from nothing in 6 days by supernatural means, that no species have arisen since that time of creation, and that fossils are the traces of creatures trapped during a recent global flood.

Why are the creationists so committed to erasing evolution from the blackboards of science classrooms? Analysts suggest that the creation–evolution debate in America has its roots in three quintessential aspects of U.S. culture: a prevalence of religious notions in politics, a core value of equal time for all points of view in public discourse, and pervasive scientific illiteracy. Behind these proximate causes is a deeper philosophical objection. Following Darwin's elucidation of natural selection as the primary force shaping evolution, no compelling justification remained for invoking the direct hand of an omnipotent God to account for life's origins and diversity, nor were there any longer grounds for assuming an unusually unique genealogical status for *Homo sapiens*. The apparent hegemony of evolutionary causation in previously sacred realms was more than many fundamentalists could tolerate.

Such views are not confined to Christian extremists. Dobzhansky opened his 1973 article by quoting from a letter to the king of Saudi Arabia from one of the country's leading sheikhs, who wrote: "The Holy Koran, the Prophet's teachings, the majority of Islamic scientists, and the actual facts all prove that the sun is running in its orbit . . . and that the earth is fixed and stable, spread out by God for his mankind. . . . Anyone who professed otherwise would utter a charge of falsehood toward God, the Koran, and the Prophet." These sentiments, of course, are strikingly reminiscent of those expressed by the Catholic Church when, in 1633, it found the scholar Galileo guilty of heresy for suggesting that Earth is neither flat nor the center of the universe.

The physical and biological sciences have given us a very different perspective on the world and its biota in space and time. No longer can humanity rationally see itself as the center of all creation, nor can we any longer rationalize continued abuse of the biosphere. Likewise, the evolutionary sciences have given us a grand temporal perspective on life. Humans have inhabited Earth for only the last few seconds of the cosmic calendar, yet already we threaten to squeeze from the planet much of the exuberant biodiversity that traces in ancestry back across four billion years. In 1949, Aldo Leopold worded the ramifications thus: "We know now what was unknown to all the preceding caravan of generations: that men are only fellow-voyagers with other creatures in the odyssey of evolution. This new knowledge should have given us, by this time, a sense of kinship with fellow-creatures; a wish to live and let live; a sense of wonder over the magnitude and duration of the biotic enterprise."

I have argued that an expanded literacy in the evolutionary and genetic sciences will be crucial in the coming decades if societies are to address the growing technical challenges in biology. Nonetheless, I suggest that with respect to the most critical and urgent challenge of all—shepherding Earth's biodiversity through this critical bottleneck century—science and religion can and must put aside their philosophical differences, at least for now, and join forces in a crusade to save the planet.

Organized churches and religious leaders have a tremendous opportunity to play key moral as well as orchestrational roles. For example, the Koran encourages Muslims to examine the beauty of their natural surroundings with curiosity and attentiveness. The Bible encourages Christians to shepherd God's creation. The teachings of Buddhism emphasize personal ethics and wise restraint. Indeed, every major world religion encourages appreciation and respect for our surroundings and for life. Organized religions can offer both moral authority for preserving biodiversity and logistical expertise to convey that imperative to their congregations. Yet sadly, most religious leaders have spurned this calling.

In his 1973 article, Dobzhansky wrote: "I am a creationist *and* an evolutionist. Evolution is God's, or Nature's method of creation." Naturalists of earlier times typically were deeply spiritual also. Consider the natural theologians. When William Bartram roamed the southeastern United States 200 years ago, he sought through his naturalist studies to glorify the works of God. On nearly every page of his diary, Bartram expressed a sense of wonderment: "This world, as a glorious apartment of the boundless palace of the sovereign Creator, is furnished with an infinite variety of animated scenes, inexpressibly beautiful and pleasing, equally free to the inspection and enjoyment of all his creatures." When John Muir, another famous natural theologian, explored the western United States a century later, he often wrote of God's bounty as well: "The charms of these mountains are beyond all common reason, unexplainable and mysterious as life itself . . . I thank God for this glimpse of it."

When Darwin set out on his voyage of discovery in the early 1800s, he was a natural theologian seeking to understand the nature of creation. He had no idea that his discoveries soon would revolutionize rational thought about nature and humanity's place within it. Yet, like Bartram and Muir, he also retained a sense of nature's magnificence, as illustrated by the famous closing paragraph in *On the Origin of Species:* "It is interesting to contemplate a tangled bank, clothed with plants of many kinds, with birds singing on the bushes, with various insects flitting about, and with worms crawling through the damp earth, and to reflect that these elaborately constructed forms, so different from each other, and dependent upon each other in so complex a manner, have all been produced by laws acting around us. . . . There is grandeur in this view of life, with its several powers, having been originally breathed by the Creator into a few forms or into one."

In the twenty-first century, an urgent challenge will be to instill in our collective psyche a moral commitment to Earth and its remaining biodiversity. Like the natural theologians and the early evolutionists, societies must find a way to integrate the emotive power of religion with the rational insights of science. They could thus promote a deeper respect for nature and engender a passion to preserve it.

With the grave responsibility to protect biodiversity from human impacts also comes a magnificent opportunity. Consider a future world in which human societies universally seek to attain a sustainable relationship with nature. Imagine that this planetary ethos eventually becomes so deeply ingrained within our collective psyche that it becomes one of mankind's defining legacies. Although the shift would be a huge, it is within the realm of possibility. Perhaps the sciences of evolutionary biology, genetics, and ecology (hopefully with encouragement from organized religions) can help point the way toward such a new environmental ethic. Relevant words were spoken by the late pope, John Paul II (who also famously told Catholics that evolution seemed logical to him): "In our day, there is a growing awareness that world peace is threatened not only by the arms race, regional conflicts and continued injustices among peoples and nations, but also by a lack of due respect for nature, by the plundering of natural resources and by a progressive decline in the quality of life." Perhaps with such words of chastisement, humanity can be encouraged to rise to a grand opportunity to save the biosphere, and with it ourselves.

16

Models, Metaphors, and Machines

This chapter offers another view on metaphors in science.

Silly Putty without the word *silly* was just mundane putty until 1949, when an unemployed ad man named Peter Hodgson added bright colors to the substance and packaged it in plastic eggs. Silly Putty became one of the most successful novelty toys of the twentieth century, grossing hundreds of millions of dollars in sales and achieving name recognition in more than 95% of American households. Evocative metaphors in science are quite like Silly Putty: colorful, alluring, simply packaged, widely recognized, enduring, and highly moldable. However, they also have a more serious side.

As models, scientific metaphors can be visual, conceptual, linguistic, digital, mechanical, or mathematical interpretations that encapsulate complex phenomena in oft-brilliant images intended to evoke an enhanced sense of cognition or understanding by the finite human mind. But what constitutes understanding? What counts as knowledge? As explanation? As progress in science? What are the ways of knowing? What do scientists want, and how well do metaphors, models, and machines help to meet those desires? Indeed, what is reality as compared with facsimile, analogy, correspondence, or similitude, and to what extent do the lines blur as metaphorical models improve? Such are the provocative intellectual and epistemological topics that Evelyn Fox Keller eloquently addresses in this insightful book.

The theatrical stage for Keller's treatment is the historical study of animal embryogenesis. Trained as a theoretical physicist, Keller had been indoctrinated in the notion that mathematics and logic provide determining arguments, and that experiential evidence, by contrast, is fallible. Later in life, when she moved into molecular biology, she was aghast to discover that most developmental biologists view experiments and observations as far surer paths to truth. Several sources underlie the differing outlooks in these fields, but an early revelation for Keller was that many biologists interpret the supposed benefactions of mathematical models—their economy of explanation, focus on idealized settings, simplifying assumptions, and pursuit of the theoretically imaginable—as weaknesses rather than strengths. Although physicists had canonized these attributes as cardinal virtues, developmental biologists often interpreted mathematical descriptions as pointless window-dressing—perhaps elegant in construct, but ultimately sterile for yielding genuine insights into the nature of complex living beings.

Making Sense of Life is an effort to understand what qualifies as scientific knowledge, using the field of embryogenesis as the touchstone. Divided into three parts, the book traces major empirical and conceptual episodes in the history of developmental biology during the twentieth century, and explores the changing landscape of what has and what has not counted as scientific enlightenment in that discipline.

The first part of Keller's book summarizes the nongenetic orientation of the first several decades by comparing three scientists' extended metaphors of cellular operations and organismal development. Stéphane Leduc's physical models entailed manipulations of inorganic chemicals (e.g., potassium nitrate on glass slides) and material processes (e.g., osmosis and diffusion) to mirror, as artificial fabrications, biological phenomena such as cell division or the emergence of lifelike morphological structures. D'Arcy Thompson's graphical models and Alan Turing's mathematical representations likewise were developed with the broad intent of illuminating developmental processes. Each of these scientists believed his approach to be of great significance in interpreting the ontogeny of form in the real biological world. Indeed, some of the representations were so mimetic of living operations as to be deemed essentially equitable with them. Keller closely examines these metaphorical constructs in the context of their own times and from the perspective of today. She critically asks, for example, why history made an icon of Thompson and an embarrassment of Leduc, when both developed provocative new ways of looking at the problem of individual development.

The second part of Keller's book turns to more familiar models of development that emerged from molecular biology and genetics during the latter half of the twentieth century. Terms like *gene action, genetic program, positional information, genetic code, DNA blueprint, developmental switches, gene batteries, networks, steering mechanisms, metabolic feedback, pattern formation,* and others are lexical devices that paint metaphorical concepts of cellular workings and ontogeny. Keller asks "What's in a name?" and answers that a serviceable language metaphor summarizes existing ideas in simple recognizable terms, stimulates further scientific inquiry, and even becomes, in a deeper sense, our actual understanding of reality.

Keller also notes that some of the best metaphoric constructs are little more than surrogates for ignorance, giving an illusion of truth yet remaining ambiguous in the lack of genuine understanding. Being initially amorphous, like putty, they can change shapes (meanings) with time to accommodate new scientific discoveries. Indeed, the most successful metaphors become immortalized by death, incorporated so fully into conventional wisdom as to avoid the radar of consciousness. Camouflaged by their own banality, we no longer notice such metaphors for what they are, and think them the actual truth.

Part 3 of her book discusses turn-of-the-century developments in recombinant DNA and computing technologies that bear on the notion of what counts as useful explanation in developmental biology. One of Keller's themes is that the manipulative capabilities of physical genetic engineering and the new powers of computer modeling have some epistemological parallels with twentieth-century approaches in developmental biology and theoretical physics. The new initiatives, Keller suggests, some day may even facilitate a greater rapport between these two seemingly different kinds of scientific culture.

In the end, however, Keller concludes that the human mind, being itself a part of the natural world and having evolved for reasons largely unrelated to making scientific sense of things, may never be able to grasp in rational terms all that we encounter. This will be especially true, she argues, when real phenomena are as complex as embryonic development would seem to be. She endorses instead an "explanatory pluralism" that not only accepts differences in epistemological cultures but is itself a virtue, representing our best chance to come to grips with the natural world. Keller contends that "A description of a phenomenon counts as an explanation . . . if and only if it meets the needs of an individual or community," and that "The question of what qualifies as a scientific explanation may not be answerable in absolute terms, but perhaps . . . in terms of the particular human needs that are, after all, the *raison d'être* of the entire pursuit."

Personally, I do not fully share these relativistic notions about science, nor their rather gloomy postmodernistic ramifications about the limits to genuine understanding in biology. Nonetheless, Keller's work is intelligent and I recommend it highly. The book provides an informative history of developmental biology, and an even more penetrating look at the epistemology of science. Not least, the book should serve as a powerful antidote to any egoistic scientist or self-congratulatory biological discipline that takes its metaphors too seriously. Life is, after all, nothing more nor less than evolution's silly putty.

17

Evolution's Unanswered Questions

We have come far in recent centuries and decades toward under-standing how life evolves. Yet many outstanding questions remain, and undoubtedly countless more are not yet even imagined. Every evolutionary biologist must have a list of favorite scientific questions, and here are four of mine.

1. What are the evolutionary bases of animal behaviors and instincts?

Especially in recent years, scientists have come a long way toward de-ciphering the genetic and developmental underpinnings of many morphological, physiological, and biochemical traits in a wide range of species. Far more refractory, it seems, are attempts to understand the precise mechanistic bases of animal behaviors and instincts. How in the world do genes program into a honeybee the capacity to com-municate to a hive, through details of its dance, the distance and di-rection of a nectar source? How do segments of DNA dispose and en-able a female sea turtle to navigate thousands of kilometers of ocean in a return migration, after an absence of more than a decade, to nest at her natal site? How in the world do physical pieces of genetic mate-rial instill an orb-weaving spider or a hummingbird with the drive and ability to craft intricate webs of silken strands and lovely nests of lichens, respectively?

An amazing but indisputable empirical fact is that such instincts and behaviors gradually evolve in various lineages, just as other cate-gories of organismal phenotype do. Thus, heritable genetic variation for such traits must be present. A major challenge for evolutionary bi-ology is to illuminate, in particular instances, precisely how informa-tion stored and transmitted in physical molecules (of DNA) can ultimately translate into such seemingly ethereal outcomes as "instinctual knowledge."

2. Why did sleep evolve?

At face value, sleep might seem to be highly maladaptive, wasting precious time that otherwise could be spent in fitness-enhancing activities such as foraging, finding mates, and rearing offspring. By regularly disabling a creature's senses and awareness, sleep also increases risks to predation, and in general diminishes an animal's capacity to respond properly to immediate threats or opportunities in its environment. Of course, the pat medical response is that sleep is necessary for energetic rejuvenation, but the deeper question remains—why? In theory, the behavioral tactic "sleep" should be evolutionarily fragile, because any genes that predispose animals to get by on less sleep should enjoy a large selective advantage over their sleep-inducing counterparts. If natural selection for sleeplessness is indeed strong, why the phenotype has seldom or never evolved presents an enigma.

The paradox of sleep is emblematic of similar queries that likewise apply to many suboptimal phenotypes. Scientists understand that evolutionary traits can be slipshod, history-laden, and generally unlike what a conscious omnipotent agent might engineer. But exactly why, in each particular instance, do various organismal traits fall far short of designer perfection?

3. What is the evolutionary significance of "junk DNA"?

The genomes of complex organisms hold vast amounts of DNA that at face value appear superfluous if not detrimental to an organism's well-being. In most multicellular species, including humans, more than 50% of the nuclear genome is composed of highly repetitive elements that have proliferated (quite like viruses or miniature parasites) to oft-astounding numbers within all cells of an organismal lineage. Such elements act as if selfish, that is, vested in self-replication and evolutionary survival even if these actions negatively impact an organism's genetic fitness. Other categories of DNA such as pseudo-genes and intergenic spacer regions add to the total pool of DNA without obvious utility to the organism.

A wonderful challenge for evolutionary biologists is to gain a much better understanding of what formerly was considered "junk DNA." What fraction of this genomic trash (either fresh or recycled) will actually prove to be of near-term or longer-term evolutionary benefit to its hosts, and how so?

4. What made us human?

A draft sequence of the full three-billion-base-pair human genome was published in 2001, and the sequence of the similar-size chimpanzee genome followed just a few years later. Everything that differentiates humans from chimpanzees has evolved within the past five million years (following separation of these species from a common ancestor), and the genetic alterations responsible must reside somewhere within the relatively small fraction of nucleotides (ca. 1%) that proved to distinguish the aligned human and chimp genomes.

Sequencing these two genomes was the easy part. Now comes the demanding task of identifying and characterizing (annotating) functional alterations that might be associated with any of the approximately 30 million genetic differences (the 1% of three billion nucleotides) between humans and chimps. Which among these are responsible for the exceptional human capacity for language or tool use, for example? In other words, what are the genetic and biochemical bases of human uniqueness?

Source Articles

Publication information for the original versions of the chapters contained in this book is given below.

1 *Molecular Evolution*, ed. F. J. Ayala, 106–122 (Sunderland, MA: Sinauer) (1976)

2 *Science Progress* 64:85–94 (1977)

3 *The Florida Naturalist* 55:7–10, with Joshua Laerm (1982)

4 *Evolution* 43:1192–1208 (1989)

5 *Natural History* 98:24–27 (1989)

6 *Evolutionary Biology* 26:225–246, with Joseph M. Quattro and Robert C. Vrijenhoek (1992)

7 *Evolution* 47:1293–1301 (1993)

8 *Conservation Biology* 8:327–329 (1994)

9 *Journal of Heredity* 89:377–382 (1998)

10 *Molecular Ecology* 7:371–379 (1998)

11 *Evolution* 54:1828–1832, a book review of *Species Concepts and Phylogenetic Theory*, edited by Q. D. Wheeler and R. Meier (New York: Columbia University Press) (2000)

12 *Science* 294:86–87 (2001)

13 *Annual Review of Genetics* 36:19–45 (with 10 coauthors) (2002)

14 *Biodiversity and Conservation* 12:1–7 (2003)

15 *BioScience* 53:247–255 (2003)

16 *Perspectives in Biology and Medicine* 47:145–148, a book review of *Making Sense of Life: Explaining Biological Development with Models, Metaphors, and Machines*, by E. F. Keller (Cambridge: Harvard University Press) (2004).

Index